THE

# MODERN PRACTICE

OF

# AMERICAN MACHINISTS & ENGINEERS,

INCLUDING THE

CONSTRUCTION, APPLICATION, AND USE OF DRILLS, LATHE TOOLS, CUTTERS FOR BORING CYLINDERS AND HOLLOW WORK GENERALLY, WITH THE MOST ECONOMICAL SPEED FOR THE SAME; THE RESULTS VERIFIED BY ACTUAL PRACTICE AT THE LATHE, THE VICE, AND ON THE FLOOR.

TOGETHER

WITH WORKSHOP MANAGEMENT, ECONOMY OF MANUFACTURE, THE STEAM-ENGINE, BOILERS, GEARS, BELTING, ETC., ETC.

BY EGBERT P. WATSON,
LATE OF "THE SCIENTIFIC AMERICAN."

Illustrated by Eighty-Six Engravings.

PHILADELPHIA:
HENRY CAREY BAIRD,
INDUSTRIAL PUBLISHER, 406 WALNUT STREET.
LONDON:
TRÜBNER & CO., 60 PATERNOSTER ROW
1869.

# PREFACE.

It is not claimed by the author that this little volume covers the whole art and mystery of Machine making, but he nevertheless confidently hopes and believes that apprentices and others, who are seeking reliable information upon this subject, will here find something new and of real and permanent interest. Those persons generally who are interested in Steam-engines and Boilers, will also find various points touched upon, with more or less amplitude, which are of importance to them.

The matter will be found to be purely American, and must therefore possess additional practical value to American mechanics.

New York, *October* 13, 1866.

# CONTENTS.

## PART I.


CHAPTER I. PAGE

The Drill and its Office.......... 13

CHAPTER II.

The Drill and its Office—continued.......... 24

CHAPTER III.

The Drill and its Office—continued.......... 30


## PART II.

## LATHE WORK.


CHAPTER IV.

Speed of Cutting Tools.......... 37

CHAPTER V.

Chucking Work in Lathes.......... 44

CHAPTER VI.

Boring Tools.......... 46

CHAPTER VII.

Boring Tools—continued.......... 57

Abuses of Chucks.......... 62

## CHAPTER VIII.

PAGE
Boring Steel Cylinders and Hollow Work........ ... 64
Experiments with Tools needed..................... 67
Conservatism among Mechanics.................... 69

## CHAPTER IX.

Turning Tools.................................... 71

## CHAPTER X.

Turning Tools—continued.......................... 77

## CHAPTER XI.

Turning Tools—continued.......................... 82

## CHAPTER XII.

Turning Tools—continued.......................... 92

## CHAPTER XIII.

Turning Tools—continued.......................... 98

# PART III.

# MISCELLANEOUS TOOLS AND PROCESSES.

## CHAPTER XIV.

Learn to Forge your own Tools.................... 104
Manual Dexterity.................................. 106
Spare the Centres .................................. 108

## CHAPTER XV.

Rough Forgings.................................... 109

## CHAPTER XVI.

How to Use Callipers ............................. 113

### CHAPTER XVII.

PAGE
A Handy Tool ........................................ 115
Rimmers........................................ 116

### CHAPTER XVIII.

Keying Wheels and Shafts ........................... 119

### CHAPTER XIX.

Taps and their Construction........................... 122
Tapping Holes........................................ 125
Abuse of Files.............................. ...... 126

### CHAPTER XX.

Defective Iron Castings............................ 127
"Burning" Iron Castings.............................. 130
How to Shrink Collars on a Shaft.................. 131

### CHAPTER XXI.

Are Scraped Surfaces Indispensable?............... 132
Oil Cups........................................ 135
Drilling and Turning Glass............................ 136

### CHAPTER XXII.

Manipulation of Metals ............................. 138

---

## PART IV.

## STEAM AND THE STEAM-ENGINE.

### CHAPTER XXIII.

The Science of Steam Engineering.................... 142

### CHAPTER XXIV.

Piston Speed of Beam-engines ........................ 147

CHAPTER XXV.

PAGE
How to Set a Slide Valve........................ 153
To find the Length of the Rod.................... 155
An Improperly Set Valve........................ 157
Lead.................................... 158
The Lead Indicator.............................. 159

CHAPTER XXVI.

Defect in Steam-engines.......................... 161

CHAPTER XXVII.

The Slide Valve.............................. 164
Balanced Slide Valves........................... 167

CHAPTER XXVIII.

Connection of Slide Valves........................ 170
The Pressure on a Slide Valve...................... 173

CHAPTER XXIX.

Condensation of Steam in Long Pipes.............. 175

CHAPTER XXX.

Packing Steam Pistons.......................... 180

CHAPTER XXXI.

Pistons without Packing........................ 183

CHAPTER XXXII.

Bearing Surfaces.............................. 185

CHAPTER XXXIII.

Lubricating the Steam-engine.................... 189

CHAPTER XXXIV.

Derangement of Steam-engines.................... 194

## CHAPTER XXXV.

PAGE
Cold Weather and Steam-engines.................... 196
Oil Entering a Steam Cylinder against Pressure..... 198

## CHAPTER XXXVI.

Explosions of Steam-boilers...................... 200
Boiler Explosions................................. 201
Is your Boiler Safe ?............................. 205
Faulty Construction of Steam-boilers.............. 211
Troubles Incident to Steam-boilers................ 214
Starting Fires under Boilers...................... 217
Steam-boilers and Electricity..................... 218
Field for Improvement in Steam-boilers............ 221

## CHAPTER XXXVII.

Location of Steam Gauges and Indicators........... 225
The Laws of Expansion............................. 227

# PART V.

# GEARS, BELTING, AND MISCELLANEOUS PRACTICAL INFORMATION.

## CHAPTER XXXVIII.

Relating to Gears................................ 231

## CHAPTER XXXIX.

Leather Bands.................................... 237
Belting........................................... 240

## CHAPTER XL.

Cone Pulleys for Given Velocities................. 246
Formulæ for Cutting Screw Threads................. 248

## CHAPTER XLI.

PAGE

How to Lay up an Eight-strand Gasket............. 250
To Turn an Elbow................................. 256
Fly-wheels for Long Shafting...................... 257
Velocity of Mechanism............................ 258

## CHAPTER XLII.

Various Useful Items............................. 259

# Practice of American Machinists.

## PART I.

### CHAPTER I.

#### THE DRILL AND ITS OFFICE.

THE perfection to which metal-working has attained is one of the miracles of modern times. Tools cut iron and brass at speeds which, fifteen years ago, would have been pronounced unattainable with economy. In gun and pistol factories and in sewing machine shops the various pieces are turned, milled, sawed, planed, or ground in such quantities and with such unfailing accuracy as to command the admiration of the observer. Not only have the tools been greatly improved in their character, but the material worked upon has also undergone important modifications; by this we mean the processes to which it is subjected before it is worked by cutters. Steel is annealed so thoroughly that its character as a tough, tenacious, and stubborn metal is wholly destroyed, and it becomes as tractable, so to speak, as the softest iron. Its virtue is not destroyed by this operation, but changed, and the temper is restored again at will.

It is important to remember that these improvements in working metals were not reached by conjecture, or by a single bound, but by successive steps and careful experiment. Whatever advantages we enjoy over other nations as skilful work men is due wholly to the skill and intelligence of our artisans, and it is no hyperbole to say that they are indeed the bulwarks of the nation.

The office of the machinist's drill (one of the most important tools) is to bore a true hole of a certain size in any metal. The conditions thus imposed upon the tool are sometimes fulfilled, but oftener not, and the reasons for this are to be found in a want of knowledge on the part of the maker of a tool, and sometimes from causes beyond the control of the mechanic; for good work cannot be made with bad materials. Three-sided holes, holes crooked in the length, holes small at the top and large at the bottom, and the reverse, ridgy holes, or those which appear to have been made with a coarse-threaded tap, oblong holes, nondescript holes, compounds of each and all the bad qualities previously mentioned, are made at times by poor workmen; and as there is no effect produced in the natural world without some cause, so also may the phenomena above mentioned be traced in mechanical operations to the omission of some important point in the construction of the drill which has been overlooked, and which is essential to its perfect operation.

To drill a straight, true hole in metal of any kind, excepting lead or copper perhaps, is just as easy as to make a wretched apology which runs in every direction but the right one, and is remarkable for nothing but its unworkmanlike appearance. Neither does it take more time to make a good hole, but on the contrary, a properly made drill works much quicker and better than one badly constructed.

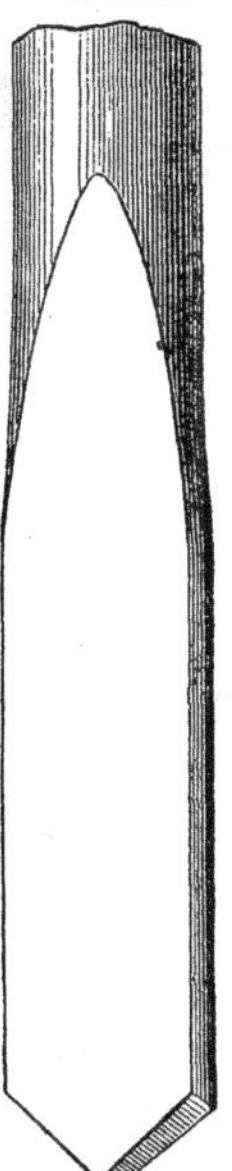
Fig. 1.

To drill a simple, straight hole in any metal we have the ordinary drill, as herewith illustrated in figure 1. This seems a very simple tool to make, but it is surprising to see what apologies for it are to be found in almost every machine shop; fig. 2 is the drill as improperly made. In the first figure it will be seen that the tool is a thin, flat steel bar for a proportionate distance, which should be so far as it is proposed to drill in the work; that the cutting edges are at right angles with each other or square, that the section shows the drill to be slightly rounded on its edges, and lastly, that the extreme point is as small and fine as it can be made consistent with strength. This is a plain, flat drill without "lips." Now the object and design of so constructing it is this: the drill should be made flat and straight, so that the borings may escape freely, and not be crushed or broken in trying to get out; neither carried round and round for several revolutions without rising to the surface, for in doing so they impede the newly formed chips below from rising out of the hole. The point should be made thin, so that it will always work true to the centre, and not tend to run out, or make a crooked hole; and the cutting edges are square for the reason that with this angle they cut equally from point to corner, and no part works faster than the other.

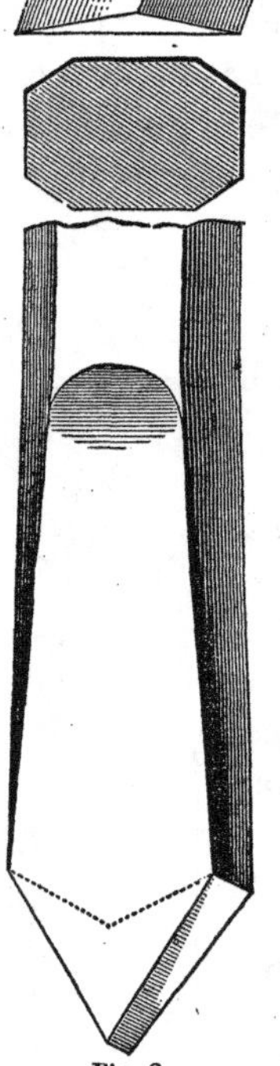
Fig. 2.

Let us take the badly-made drill, as shown in the second figure, and see what its defects are; these can readily be noticed where they depart from the well formed drill. This drill is not exaggerated in the engraving, being far short in reality of some specimens of handiwork we have seen kicking about on machine-shop floors. The dotted lines of the point and cutting edges show the various angles it is ground to, and the section and point in straight lines will now be noticed. This is a very bad drill—it is almost unnecessary to say that; it is stubbed and blunt, and could not drill an inch with decent feed without getting so hot as to draw the temper. The section is octagonal, which is the worst possible form, because the chips catch at the angles, and, not being able to get out, are ground to powder, requiring more power to turn the drill than fig. 1; the edges are sloped directly from the cutting edge to the back of the drill, and the point is thick and square. With such a drill as this a hole like fig. 3 would be made, and for these reasons: the thick point with its quick angles cuts unevenly; first the point bores down, and then the labor comes on the edges, and the drill-point works loose, making a cone center in the hole like the diagram in fig. 4; the consequence is that the hole is untrue. The sharp edges sloping so quickly toward the back are also a disadvantage, because they afford no support to the cutting edges, which go astray in consequence.

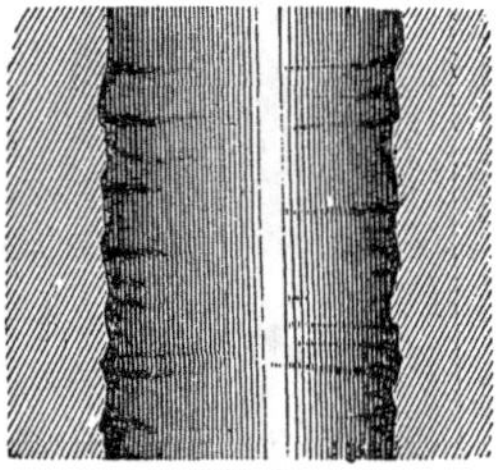

Fig. 3

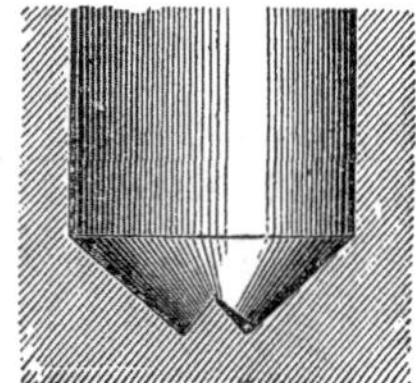

Fig. 4.

For general use a plain drill without lips is as good as any, in fact better. A "lipped" drill cuts faster but gets dull quicker, because the edges are thinner and keener, and require grinding oftener·

so it is a question whether it is any better for ordinary use. Where deep holes are to be drilled lips are an advantage, for the chips removed are heavier than the plain drill makes, and do not clog so quickly. This is a lipped drill, in fig. 5, and consists, as mechanics know, in simply making the cutting edges hollow, or thinner, so that they take a ranker hold of the metal, just as a plane iron does when pushed out too far.

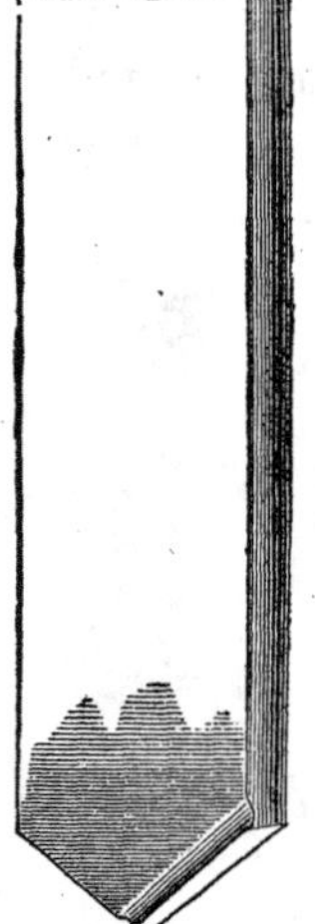

Fig. 5.

A great deal depends on grinding a drill; for while the cutting edges may be all right in shape, if they are ground at too sharp an angle, they are soon rubbed off on the work, and do not perform efficiently; or if one side is ground longer than the other the hole will not be round. The back part of the cutting edge should not be raised too high. In effect the cutters of the drill are two chisels which remove the iron as they revolve; now if we were to employ a chisel for cutting wood, the angle of inclination of the edges to the work should

Fig. 6.

be such that it would require little pressure to force them in. The tool would not be held but as in fig. 7. The force is not applied downward in this case, but in a plane, as with a screw-driver; therefore, to follow out this illustration, the drill

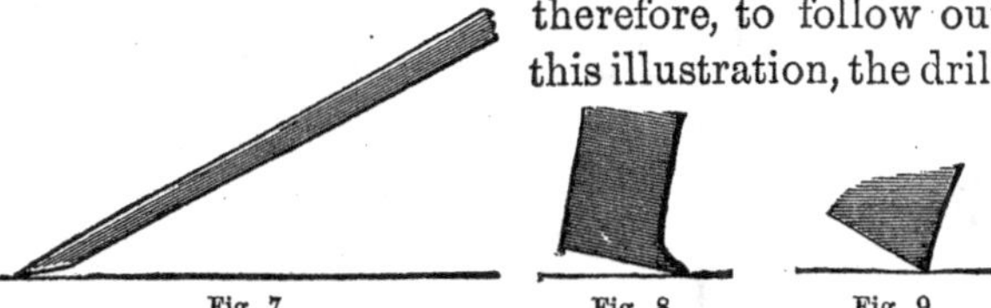

Fig. 7. Fig. 8. Fig. 9.

should have but little clearance behind, as in this diagram (fig. 8), which is merely intended to show the idea, and not as a pattern for a drill edge. Not as some grind them, in fig. 9. A planing tool also furnishes an illustration of this matter; for if a finishing flat-nosed tool was ground like the last diagram, it would do nothing but chatter, while the first would cut smoothly and without jar. These are the main points of good drills of the ordinary kind, but there are an almost endless variety of them, such as twisted, pin drills, counter-borers, etc., and each and all of these have different shapes to suit different work. It is impossible and unnecessary to go into these at length, and we shall only notice one of each kind mentioned above.

Of late, the manufacture of twist drills has been a specialty with certain parties. These drills have been thoroughly studied and experimented upon to find out their weak points. The consequence is that any mechanic can go to a hardware

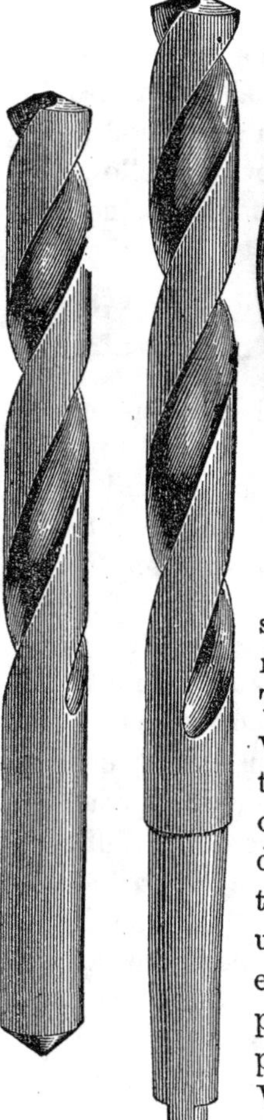

Fig. 11

Fig. 10.

store and buy a twist drill as readily as he could an auger. This is an advantage, but a very slight one compared with the fact that all these drills are of *standard* sizes, so that holes drilled by them cannot vary in the least; they are tempered up to the shank, and will cut either in steel, iron, brass, copper or *wood*, and, in fact, indispensable to all metal workers. We have for some time used sets of these drills, of both makers, and should feel lost without them. There is no occasion to alter one. All sizes are ready to the hand, and they are especially useful for tubes or holes that have to be threaded.

The sizes and styles of the Morse drill and chuck are as follows:

A set of 21 drills, with taper shanks from $\frac{3}{8}$ to $1\frac{1}{4}$ inch in diameter, embracing every size usually required by machinists. These drills are made of the best *cast steel* imported, made expressly for the purpose, and are fitted to sockets that accompany them. Three sockets are required for the set, are made either of steel or iron, as desired, and are readily fitted to any drilling machine. They are of standard size and taper, and consumers are assured that all taper shank drills hereafter made will fit them.

A set of 15 drills, styled the "Machinist's set," of sizes varying by 32ds from 1-16 to 1-2 inch, made of the *best material*, with straight shanks the same size as the drill. An adjustable chuck ($2\frac{3}{8}$ inches in diameter), capable of holding every drill, can be furnished. This is easily and cheaply fitted to any lathe.

A set of 29 drills, styled the "Jobber's set," of sizes varying by 64ths from 1-16 to 1-2 inch. This set is especially designed for trades doing fine work, as they will drill the size of the body of screws up to 1-2 inch, and are tap drills from

3-32 to 9-16. The adjustable chuck will also hold this set.

The "Wire Gauge set," comprising the sizes of Stubs' steel wire gauge, from No. 1 to 60, inclusive.

The "half set," by steel wire gauge, comprising alternate numbers, from 1 to 60; and the "Jeweller's set," consisting of 36 drills, from $\frac{1}{8}$ inch (No. 30) to No. 65 steel wire gauge.

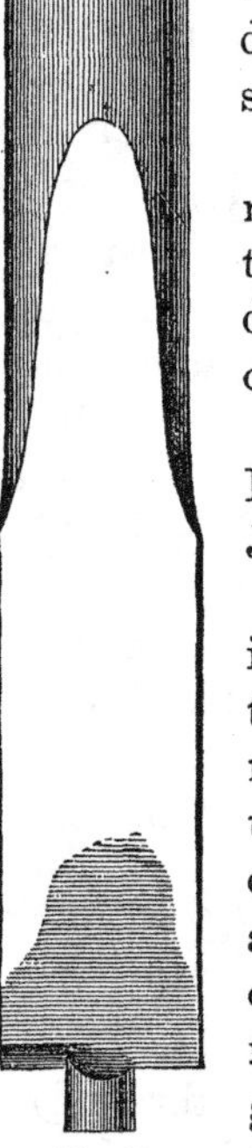

Fig. 12.

The five smaller sets are neatly mounted on stands, numbered so as to show at a glance the size of every drill, greatly facilitating the selection of the one required for use.

The other drill is the Manhattan Fire-arms Company of Newark, New Jersey, and is in wide demand.

Here is a pin drill (fig. 12), some call it a counter-borer, but this is not a term which can be applied indiscriminately, for in some jobs the tool is used wholly as a drill, and not as a cutter or tool to counter-bore, or drill against certain other holes. The use of this tool is to drill large holes more correctly and faster than a single drill could do it, and it is used

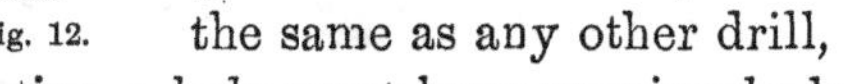

the same as any other drill, with this exception, a hole must have previously been made for the "tit" or pin, in the work to be done. If this first hole is not straight the pin drill will not go straight, for the pin follows the first hole, which is usually small, in about the same proportion as the diagram. The first hole acts merely as a guide for the pin, and when it is made true the pin drill follows in it, and takes out great curling chips of metal with the greatest facility. The pin should have very little clearance in the first hole, so that it cannot shake about, and the first hole is sometimes a trifle smaller, and the "tit" on the drill is serrated, as shown in fig. 13, so that it clears itself as it goes down, and always fits snugly. If the first hole drilled in the work is too large, the pin drill goes all over, and neither makes a round hole or a true one. These drills are costly to make, as they must be turned in the lathe and afterward filed up, since, from their conformation they cannot be ground on the stone, although they may be sharpened on a true running stone, when held by a steady hand.

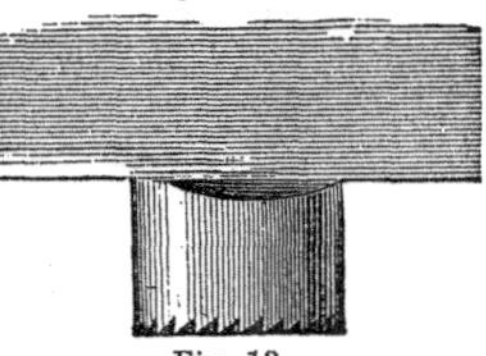

Fig. 13.

The Morse Twist Drill and Machine Company's Works are at New Bedford, Mass.

## CHAPTER II.

### THE DRILL AND ITS OFFICE—CONTINUED.

THERE is an endless variety of counter-borers or pin drills adapted to every class of work, but as the principle is the main thing, it is not necessary to follow or to illustrate every one. The counter-borer in one shape is used to cut out the tube holes in flue sheets, which in boilers as lately built require a great deal of time; if the tool is not properly made, many sizes are required in large shops, where much work is done.

Here is a plan (fig. 14) for an expanding or an adjustable tool by which holes in flue sheets can be made of any size, varying only with the plan of the cutters. The apparatus is very simple, and by altering the shape of the cutters, a hole but little larger in diameter than the rod or shaft that carries the arm can be made. The advantages of this appliance over an ordinary drill, such as is frequently used, are that the cutter, which breaks often even with the utmost care, can be easily dressed when

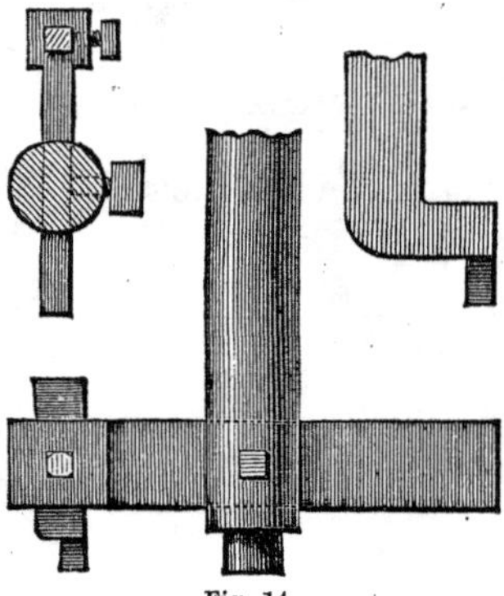

Fig. 14.

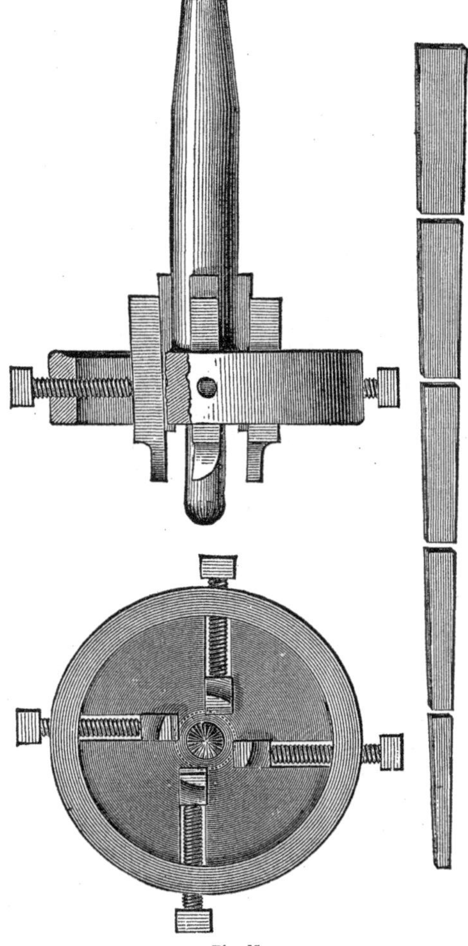

Fig. 15.

broken in much less time than the counter-borer could, thus making it cheaper to use, both in point of execution, and cost of repair when injured.

The tool in fig. 15 is, as may be seen, adjustable, and will cut holes of varying diameters, according to the sizes it is constructed for. The arrangement is simple, and consists of a central head forged solid on the shank. This head is planed out for the reception of the tools of cutters, and has, further, a wrought-iron ring shrunk over it. This ring is tapped out to receive the set screws which hold the tools fast. Behind the tools are wedges, which, when driven or slacked off, advance or retract the cutters with great nicety; the taper is planed in the shank for the wedges, so that the cutters always stand vertically. The wedges should all be planed at once, so that there will be no variation in them, and several sizes should be provided, so that holes of any diameter can be made. The cutters need not all travel in the same track, but each may set a little inside of the one that forms the size; in this way they cut freer and are less liable to break. This tool is useful, not only in the boiler shop, but also in the finishing department, for by changing the character of the cutters, work of almost any kind can be done.

Here is another plan (fig. 16) for a boring tool or tube sheet-borer more properly, for this is the object it was designed for. It is not so good a tool as the first one for some purposes, but as all persons may not have the same opinion we give it place. It is not adjustable, except limitedly, it

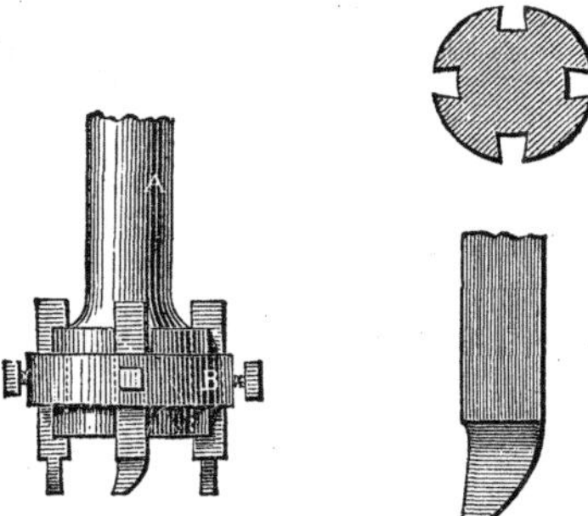


Fig. 16.

costs more to make at first, but it will work faster and do equally as good, if not better work than the ordinary adjustable cutter shown in fig. 14. The bar, A, is merely forged with a larger portion on the end, and is grooved on four sides to admit the cutters; these are simply square nosed, offset on one side, and the cutting part, of course, curved to suit the circle it works in; a wrought-iron ring, B, is then slipped over the cutters to hold them firmly in place and adjust them so that all the points may work at once; this ring has set screws for each cutter, and one of the cutters may be made to countersink the sheet at the same time, if it is preferred to do it on the side drilled from. The burrs or ragged edges left on the under side of the sheet by this tool will be very slight indeed, if it is properly made, and can be rubbed off with an old file.

Still another drill for boring tube sheets is given in fig. 17, herewith. It is one commonly used,

Fig. 17.

and is a very efficient tool when well made. It is costly to construct, however, and requires to be turned in the lathe by an experienced workman, and afterward filed up so as to cut properly. The spaces between the pin and the cutters are very troublesome to cut out in the lathe with an ordinary tool, as the work in revolving strikes square and suddenly on the lathe tool, and soon dulls it, or else breaks off the end and throws the drill out of the centres. A useful cutter for making these drills is shown in fig. 18. It is simply a steel bar turned up and bored out the size of the "tit" or pin on the drill, and has teeth cut all round the circumference, as shown below. The

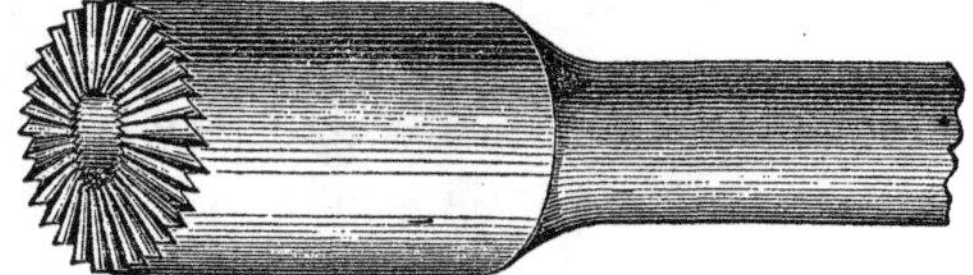

Fig. 18.

pin of the drill being slipped in the hole in this cutter, the radiating teeth cut away the central portions so difficult to remove in the lathe. The drill may revolve in the steady rest, or the barrel cutter may be so used and the work screwed up to it by inserting the centre in the dead centre of the lathe; by employing this tool much time may be saved and better work done.

A work might be specially devoted to the drill; it is one of the most indispensable of the minor instruments employed by mechanics, and it is only reasonable to add, that the tool most in use, simple though it be, deserves all the attention that can profitably be given to it.

Although the rose bit is not in any sense a drill, it is of the same class, and is indispensable to good work in the drilling machine; for if a man does not know how to grind a mill or make one, and the holes he makes are neither round, square, nor oval, then he has only to use the rose bit and he will have a perfectly round, straight hole. Fig. 19 is the bit. It may be made wholly of steel, or the

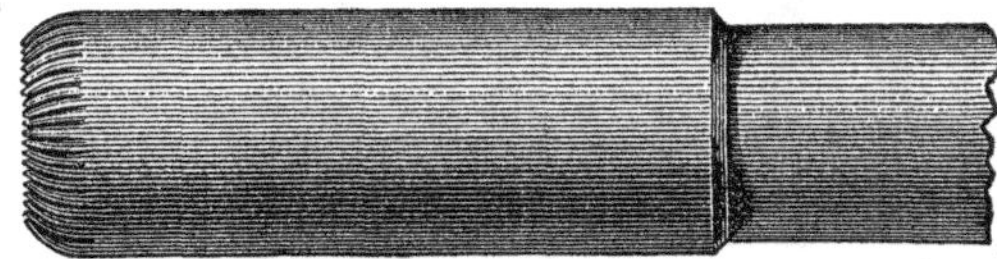

Fig. 19.

shank may be iron, and the cutting end only of steel. The end is composed of a series of fine cutters arranged regularly all around, and the body is a shade smaller at its upper end than at the lower. When the hole is drilled in the work to nearly the right size, the drill is taken out and replaced with this bit, which cuts regularly and steadily all around, and corrects any untruth in the first hole. There should be but very little metal left for it to work on, and the job must be well oiled during the process. If these conditions are observed the hole will be a true cylinder.

---

## CHAPTER III.

### THE DRILL AND ITS OFFICE—CONTINUED.

THERE is still another kind of drill for peculiar work, which is employed by some machinists, though for our own part we see no special virtue in it, for it is troublesome to use and to make, and very liable to break. It is called the tit or centre drill, and fig. 20 is an engraving of it. The centre marked out by the punch is of course the point where the tit is inserted on the work. This tit is the cause of all the trouble with the tool; it must be filed up in the vice, it tries the tool-dresser's patience to harden it, for the small quantity of metal

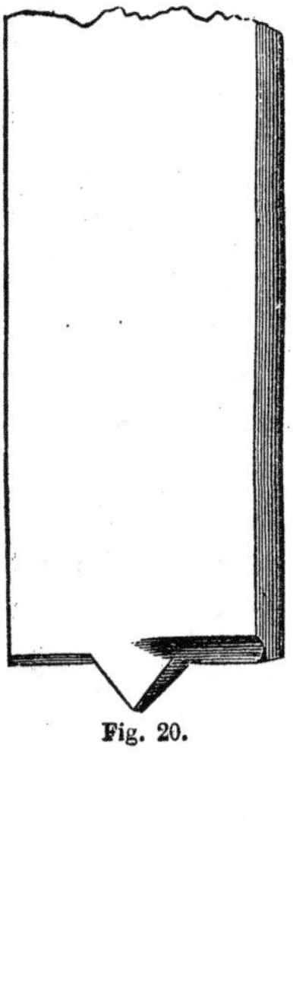

Fig. 20.

in it compared to the heavier parts in proximity, causes it to get hot in the fire more speedily and also to cool quicker, so that while the cutting edges are of the right temper the tit is soft or hard as the case may be; for all ordinary purposes the common flat drill is far superior.

Another kind of drill is illustrated in fig. 21; it is a turned drill, and will go, if it runs true in the machine, as straight as a die in the work. These two figures are side and end views; the tool is simply forged and then turned up in the lathe afterward, and it is much used for drilling holes in the tube sheets of surface condensers. Composite drills are those made by combining cutters with drills in such a manner that while the hole is being drilled, or just after the operation, it is also countersunk on top, or counterbored to a certain depth; and this without removing the drill from the hole, thus saving a great deal of time. When the tube sheets of surface condensers are drilled, such tools do good service, for the vast number of holes requires some such method to render it economical, as well as to expedite the job. The plans for a drill capable of be-

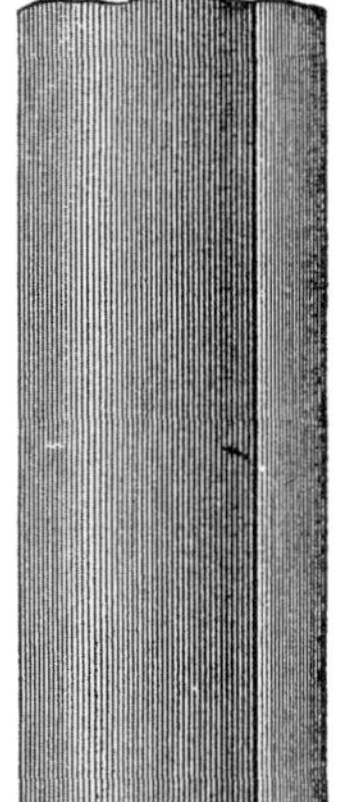

Fig. 21.

ing used for such work are given in fig. 22. The drill is simply a turned steel bar, flattened on the end for but a short distance; as the plate to be drilled is not thick, it does not require to be long, but should be made as short as possible. There is a key-way or slot, in the shank in which the cutters are set, and secured by a small key at the back. The shape of the cutter fitted in the key-way, of course varies with the work to be done, and the corners may be rounded off to make a round bottomed hole, or made to conform to any pattern desired, and the key may be made short, so that the cutter can go clean through. Drills of this kind are also extremely useful for counter-boring in lathes; a dog may be slipped over the round shank and screwed up while the centre in the drill shank is received by the dead centre of the lathe. It is much more economical to use a tool of this kind, where the circumstances admit of it, than to bother with boring tools of the usual pattern. It is in the minor details of this kind that workshop economy may be practiced to advantage, and there

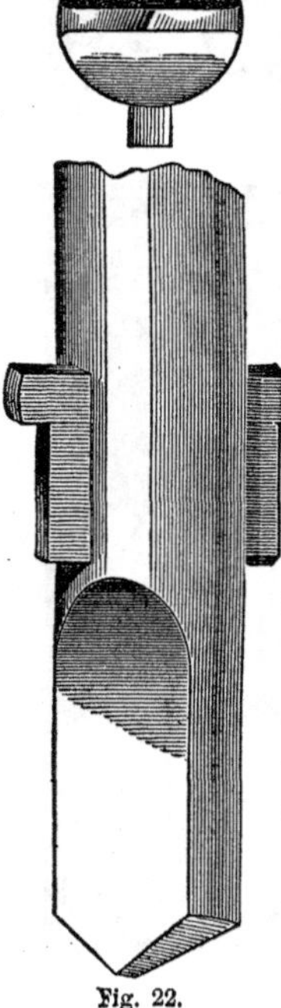

Fig. 22.

is nothing that calls more for the exercise of ingenuity than the simple matter of drilling holes speedily and accurately. In every instance it must be borne in mind that it is of the utmost consequence that the drill should run *true* on its end. Without this the finest temper and the best shape are of no value, and it is impossible to do good work where the point of the drill describes a circle of greater or less diameter.

From specific designs of drills let us depart at present, and turn our attention to the other end of the same tool, where we shall find something worthy of attention. We might fill page after page with drills of peculiar shapes; those with and without lips, those with lips or cutting edges curved, so that a section would be like this, ɷ; others with round corners, etc.; but as the main principles of drills have already been given, it is not necessary that we should follow out every design, as it would interfere with more important matters. Let us look at the drill shank. It is a common and a favorite expression with many that the minor trials of life cause more sharp annoyance and vexation than severe visitations. Be this as it may, it is very certain that the simple matter of the formation of the drill shank has caused more profanity, delay, and actual pecuniary loss than any similar part of any other tool. The shank is in general made square and taper, as in fig. 23, and the adherence to this form, the most injudicious and expensive that could be devised, is remarkable. Drilling machines upon new plans are made every day, and are fitted with some in-

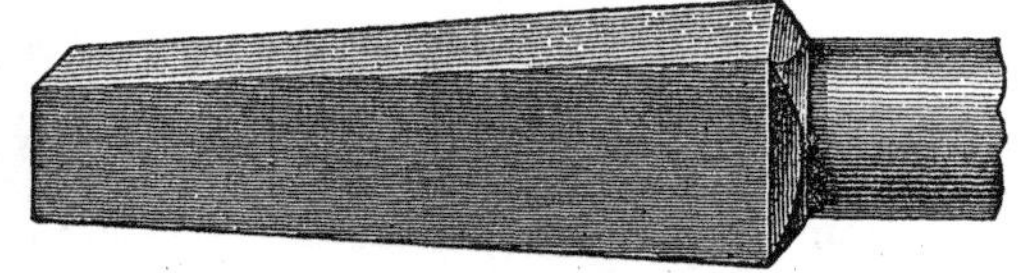

Fig. 23.

genious device for expediting the work, but for some inexplicable reason the spindle is squared out, duly tapered, and with—the height of absurdity—a set screw in addition. It is among the impossibilities of mechanical practice that a square-shanked drill should ever run true by any possibility, except one involving great expenditure of time and consequently money. It must be acknowledged by every unprejudiced person, that the true shape for a drill shank is round and parallel, not tapered like a lathe centre. With this form the drill in all cases will run much truer than with any other shape; not only is this assertion correct, but the labor or cost of making the drill shank in this form is not to be mentioned with a square or taper one. The round hole in the spindle of the machine is capable of being wholly finished in the lathe, so that when it leaves that tool it is completed, and does not require to be chipped out or even filed. Squaring the hole makes it untrue with the centre of the spindle, even when great care is used, and the drills themselves have to be forged exactly alike or else they will not fit. In a shop where there are thirty or forty drilling machines and a thousand drills there are scarcely any two alike, and when a square-shanked drill is put into a squared spindle, the point describes a circle of no small magnitude. Then comes the corrector of this evil—bang goes the hammer—the drill falls out, and a piece of emery cloth is wrapped about it because it is rough and holds better; the tool is replaced, and the same process goes on again and again; sometimes varied by breaking the drill short off at the shank, at others only succeeding, after much time and trouble, in making the drill run true. Each time it is dressed the drill is altered, so that it is no exaggeration to say that it never runs twice alike. The set screw is a nuisance, it is of no use at all; when

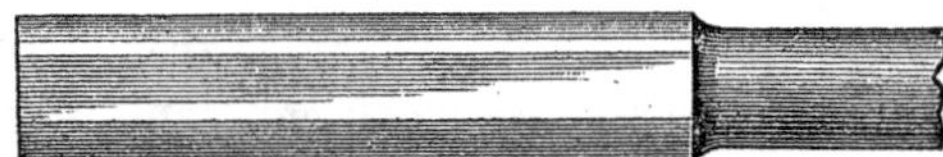

Fig. 24.

set up to its place it strikes one-sided, and instead of securing the drill actually pushes it out. How easy it would be to avoid all this complexity by making the shank in this form, or forging the drill of round steel! There are many advantages in this, although round steel is not uniformly of as good a quality as square steel. The most marked advantages are lessened first cost of construction, greater efficiency of the tool itself, and less time expended in straightening and setting the drill; a standard size for all drills, so that each one will fit every machine in the shop, and less work in making the drill machine itself. The taper round shank drill is not so good, for these reasons: it costs more than either of the others, it is troublesome to get out of the machine, for a key has to be driven in at the end, which often gets lost. The hammer is used to loosen the drill, by men too lazy to take the key when it is not lost; the taper gets bruised by the blacksmith in dressing the drill; when the drill has to be upset, as it does at times, the taper is injured on the end, and don't fit without filing; and lastly, it cannot be extended as the straight round-shanked drill can. By this we mean that sometimes a drill is just an eighth or one fourth too short to go through the work with all the screw that can be got. If a taper round-shanked drill is used, the workman must either get another, or else derange his work, to block it up higher; but if we have a straight shank we may put a piece of round iron in the spindle, and let the end of the shank bring up against it, and thus attain the end with but little trouble. Thus the straight round shank appears to have decided advantages over any of the other plans.

# PART II.

## LATHE WORK.

---

## CHAPTER IV.

### SPEED OF CUTTING TOOLS.

In making estimates for steam or other machinery, it is important to know the time required to execute certain portions of the work. This can only be accurately arrived at by a knowledge of the speeds at which cutting tools work, for mere manual dexterity cannot always be counted on for the speedy accomplishment of a job. Where of old, hammers clinked incessantly on chisel heads, and a mighty force of "chippers and filers" were busy in hewing out a cross-head or cross-tail from the rough shape in which it left the forge, a sturdy tool now cleaves its way through the tough metal, reeking with steam and bent on accomplishing its purpose, no matter how heavy the cut.

The speed of cutting tools which are employed in the largest shops, and by the most enterprising builders of machinery, also affords a standard of comparison for others desirous of excelling; and much good may be accomplished by recording the results of our observations on this subject; in this article none but metal-working cutters are considered. There are some generic difficulties in the way of arriving at positive statements concerning the rates at which cutters should travel, or the metal which is worked should pass under them, as it does in some cases; nevertheless it is sufficient to notice the speeds generally employed, and leave particular cases to take care of themselves.

In running lathes, the kind of iron, the nature of the work, the stage of the work, whether finished or merely roughed out, the texture of the metal, and the kind, whether steel, brass, cast or wrought-iron, deserves attention; for without such consideration, statements are worth no more than mere speculation, and tend to confusion. Wrought-iron and cast-iron, in small sizes, where both metals are of good quality, sound and even, free from sand streaks in the first, and flaws and blow holes, in the second, may be run at nearly equal speeds with economy (that is for all pieces under one inch and a quarter diameter, or thereabouts); as the diameter increases, however, this is no longer the fact, and cast-iron must run slower or the tool and work will be destroyed. In shafts of 3½ inches diameter, a circumferential speed of the work, equal to 220 inches per minute, or twenty turns with a lateral feed of one thirty-second part of an inch, reducing the work three eighths of an inch, is a good speed for short lengths, and one that cannot probably be exceeded with economy. Of course, in taking a cut of three sixteenths of an inch, the work would be roughed out, and the labor on the tool much greater than with small cuts and finer feed. A great deal of misapprehension is manifested on the subject of feed, many persons supposing that one thirty-second of an inch produces a chip of that thickness; this is not the case; the cut is much thicker than the feed employed, and no comparison can be formed, from the thickness of the chip, what the actual feed or depth of the cut was at each revolution. This is on account of the corrugation or upsetting of the metal as it is taken off the work by the tool, thus adding to its size. At the rate above spoken of, the tool, in turning a shaft 3½ inches in diameter, travels over a circumferential surface of 13 feet 3 inches, supposing the feed to be thrown out of gear; each revolution of the shaft advances the tool the thirty-second part of an inch; therefore a cylinder twenty thirty-seconds or five eighths of an inch long by 3½ inches diameter, is travelled by the tool in doing its work; the convex surface of such a cylinder is about 13 feet 6 inches.

At this speed, the tool, when properly dressed, ground, and tempered, works economically; taking a good cut, retaining its edge without too frequent grinding, and making smooth work. In turning wrought-iron shafts as large as 12 inches and 18 inches diameter, four and a half to six revolutions per minute, with a feed corresponding, is a "good to fair" speed, which probably could not be increased without incurring loss of time in dressing and grinding tools. This would give a circumferential velocity of about 19 feet per minute for the 12-inch shaft—6 feet more per minute than cast-iron work should be driven. Wrought-iron will bear to run quicker than cast-iron, chiefly because it is freer and softer to cut than cast-iron as a general thing, and also for the reason that the tool is generally kept cool by a stream of water, which it is not convenient or necessary to use on cast-iron.

Some planers that we timed in different shops run as follows:—A thirteen-foot bed, with a moderately good roughing feed, cuts 15 feet per minute; it must be borne in mind, however, that planing machines have to work all kinds of metal on the same speed. This has always been a cause for wonder with us; why not put one pulley on a lathe, and drive that at one speed for large and small work? A planer has exactly the same duty to do in reality as a lathe, the increasing or decreasing diameter of the work in the lathe being fully compensated for by long and short jobs on the planer. Of course a change of speed is not required in extra large tools, where the work is always heavy; but for planing machines with beds 12 feet long, there is no reason why there should not be two speeds provided on the counter shaft overhead which drives the back-and-forth motion belts. It would be just as appropriate to have only one feed for all work, as only one speed for all kinds of jobs. While a planer with a bed 12 feet long is able to run at the rate of 15 feet per minute, a 25-foot bed planer is driven at 13 feet per minute, or a little less than the first mentioned tool of this class, because heavier work is done upon it. A shaper, or quick working planing machine for short jobs, has the change of speed we speak of, and at the slowest speed, with a seven-inch stroke, also runs at the rate of 12 feet per minute. It will therefore be seen, from the examples mentioned previously in this article, that the average speed at which cutting tools work is, on work of an ordinarily heavy class and of a standard kind, about 13 feet per minute for cast-iron, and 15 feet per minute for wrought-iron. Brass we have not considered at all; but with free-working yellow brass, 25 feet per minute is not an unusual speed for diameters of ten inches. The speed of brass work, however, is a very difficult thing to fix, for while a large brass ring may run at a rapid rate and turn economically, a shaft of the same size could not be run at half the speed the ring does and work well on the tool. Besides, this brass is such a "fickle" metal, so to speak, being at one time hard and another soft, now free from "blow-holes" and again honey-combed with them, each and all phases requiring different speeds to suit circumstances; for these reasons we say we have omitted any discussion of the speeds proper to cut brass. And it must be also borne in mind that the rates mentioned as proper for cutting cast and wrought-iron vary greatly at times. Of course chilled iron cannot be turned on speed that would be proper for a soft loam or a green sand casting neither can a scrap iron shaft be run economically at the same rate as a rolled one of similar size. But as a standard rule for ordinary work, 13 feet per minute for cast-iron turned in lathes, 15 feet for turned wrought-iron, and 15 feet for planers, alike on all metals, is a fair estimate. It is very possible that our readers may have some experience which conflicts with these statements: if so, we would be pleased to hear from them.

The practice of knocking off the centres of turned work is a mischievous one. It is merely doing work that is not only needless, but that at some future day will have to be done over again. When a centre is once properly made in a shaft, or any other part, it is unalterable except by chipping or purposely changing its position, and work once turned true on good centres will always be true, provided no damage occurs to it. It is just in this particular that the true centre is useful, for if a shaft is bent or an arm on one thrown out of line, the old centres are available, and the injured piece can be made as good as new in a short time. Suppose, however, that the journal of a shaft is worn oval, or that the collar is battered and jammed up, how is it possible to find the true centre of the shaft? It never can be found; the shaft may be made to run straight, but not by its old centres if they have once been cut off. When shafts are forged too long, in cutting them to the right length great "tits" are left on the ends, which are both ungainly and in the way. This is the blacksmith's fault, and must be remedied by the machinist; cut the shaft to the right length first, knock off the centres if they are too long, and then re-centre the job and finish it according to the drawing. In steam-engine work especially, the centres of shafts are essential to nice adjustment, and they should never be removed.

A foolish notion prevails among some mechanics that centres injure the finished appearance of the work, but it seems to us that this is an erroneous view, which ought not to be tolerated. Drill every centre, and drill it deep; countersink it so that it will have a good bearing on the centres of the lathe, and the workman will have the satisfaction of knowing that, all other things being equal, he will have a good job, and one that can at any time be easily repaired if damaged.

## CHAPTER V.

### CHUCKING WORK IN LATHES.

ONE of the most indispensable adjuncts of a lathe is a chuck for holding work that cannot be turned between the centres, or requires to be bored out. Very great ingenuity has been displayed in constructing chucks so that the piece held, if round, should run perfectly true without any further adjustment. To this class of chucks belong the scroll, the worm and spiral gear chuck, and others; their utility is very great, and on some work they are indispensable.

Ordinary chucks have four jaws, which slide in grooves in the face plate, and are set up by screws running through them. Such a chuck plate can be altered to take an irregular form, or one that has a hole out of the centre, as an eccentric, but the scroll chuck cannot. The jaws in this move arbitrarily, or toward the centre, and are therefore unchangeable; although we believe there is one variety of scroll chuck in the market that can be shifted so as to take an irregular form. It is surprising to see what clumsy work some men make in chucking a job. To set a simple pulley takes them half an hour, and at the end of that time the face is so covered with chalk marks that it looks as if it were whitewashed; hammer marks indent the work, and the workman loses his patience and gets out of temper for nothing. It is the simplest thing in the world to set a round job true in a few minutes, and without chalk, sticks, or any other aid. When a pulley is to be bored, the centre should be put in the spindle, and the size measured off to the chucks; one of them should be drawn out a little to let the work in, and when it is in place, setting this slack jaw up will bring the pulley fair. One or two revolutions of the lathe will show in a moment if the outside is true. It is unnecessary to tell the mechanic that no work must be turned from the hole cored out rough. Many unthinking persons have done this to their own and the proprietor's sorrow; the cores not unfrequently get pushed on one side in casting, which makes the work all wrong if they be taken as the centre.

Scroll chucks, in fact chucks of any kind, are costly tools, and not within the reach of every mechanic. To such, a common block of wood is by no means a useless thing. It is astonishing how much can be done in a wooden chuck when properly made. Very large sizes can be employed, and for very small work, it is unequalled as a substitute for the metal chucks. Very frequently cements, such as gum-shellac, etc., are used in connection with the wooden chuck to hold small flat pieces that have no flange or other point to catch. An eccentric may be bored for the shaft and turned outside in a wooden chuck, or on a face plate, without the use of a chuck at all.

In cases of irregularly-shaped jobs, where it is at all practicable, the chuck plate should be taken off and laid on the bench, and the work set true upon it in that position; by the aid of the lines which are struck, or should be, on every plate, this can be done much more quickly than when the work is hanging by one or more bolts. In all cases the plate should be carefully used and cleaned when done with, not left to knock about on the floor under the lathe, or to get filled with grease, dirt, and chips.

## CHAPTER VI.

### BORING TOOLS.

THERE are two specific classes of tools for cutting metals. These are roughing and finishing tools. Others for different purposes, such as scraping, forming by pressure, and manifold uses, cannot properly be included under the head of cutting tools. To simplify this article we have considered the machinist's boring tools as divided into two kinds only, those for roughing and those for finishing.

A small hole can be more quickly made with a

good drill than with any other instrument, but this tool is only available for ordinary work. When we come to more exact and complicated jobs, the lathe must be used instead of the drill machine, and the boring tool, in one form or another, supplant the drill. With all roughing tools the object is to remove as much iron as possible in the shortest space of time with economy. The question of economy is not confined to merely driving the tool through the hole quickly, but also relates to the number of times the workman is obliged to go to the stone to renew the edge, or to the tool-dresser to have the same drawn down or tempered. If it be admitted that the fibres of wrought-iron, or the crystals of cast metals, must be cut and not abraded in working them, it is evident that there is but one mechanical power that will do this. That one is the wedge. To the wedge then is due all of the credit in accomplishing the object in question, but on the workman rests the responsibility of so placing the wedge that it will work to the best advantage. In this point lies all the difference between a good and a bad tool. This assertion must be strengthened by the supposition that the quality of the steel of both tools is the same, and the workmanship identical in all other respects than the shape of the cutting edges. In one position the wedge cleaves particles asunder, in another it abrades, or does its work by scraping. These qualities are shown in the annexed diagrams, figs. 25 and 26. It is not claimed as any original discovery of our

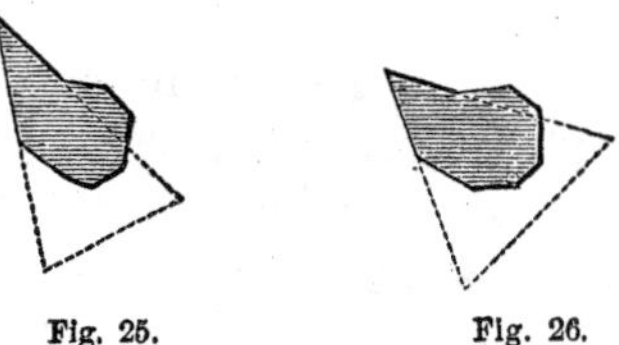

Fig. 25. Fig. 26.

own, but is only presented as a palpable and acknowledged truth among good mechanics.

In fig. 25 we have a mere sectional elevation of a common boring tool. The dotted lines show the direction of the acute end of the wedge; fig. 26 is a scraping boring tool, which also shows the application of the principle alluded to. Very often the improper application or construction of boring tools makes the hole bored out taper, or small on the back end. The unskilful workman charges the lathe with the difficulty when the fault is often his own. A good boring tool will cut free, soft metal just as well inside of the hole as a turning tool properly made will out of it, but there are very many who are content to look on and see a boring tool grate a few miserable scraps of iron out of a hole. The process bears the same relation to cutting that rasping on a grater does to shaving with a razor.

Of these tools—respectively figs. 27, 28, and 29—the mechanic will readily select the one which will cut the best, and on all metals, except brass, do the most work. The round, acute, sloping edge draws into the work, instead of springing

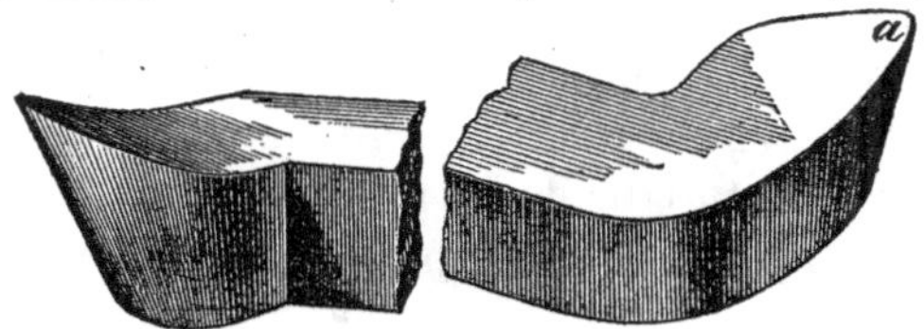


Fig. 27. Fig. 28.

from it, and holds on to the metal, producing in wrought-iron and tough brass, long curling chips that leave the tool with but little heat or compression. Some workmen, we are sorry to say, are so shiftless or indifferent, that they would take the badly-made tool in preference to the proper one, fig. 27. If a man be judged by the company he keeps, a machinist may be rated by the character of his tools, and his work will show faithfully whether they be of a proper or improper shape. The shank of fig. 27 is made square, or as nearly so as possible. In that form it is stiffer than in any other, and the only rounded part of it is the angle furthest from the edge. This is rounded to

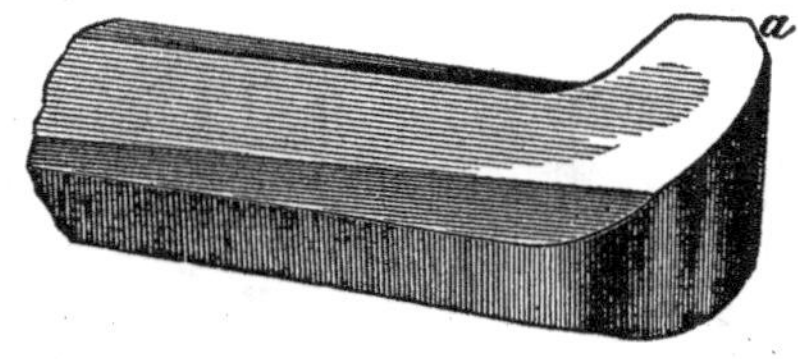


Fig. 29.

clear the work, for sometimes when a boring tool rubs at this point the cutting edge is forced in, the size enlarged, and the job spoiled. This is not the case with the other tool, fig. 29. It is one of a class in common use, and is not well adapted to the work required of it. An angle at the cutting edge (as at *a*), tends to force the tool off its cut, and to make the hole taper, as explained in a previous paragraph. The strength of the shank is lessened by being made octagonal, and the clearance in front is so slight that the tool often rubs at this point and produces a bad surface. The chips from it are stiff and corrugated, and look as if they were (as they are) ground out instead of being cut, and the whole form is objectionable. How much pleasanter it is to work with a tool like that shown in figs. 27 and 28; to have it well tempered, dressed, and sharpened, and to drive it through the hole as fast as the nature of the work will allow it to go.

The engravings just considered represent roughing tools of one kind, and are intended chiefly to bore out the heaviest portion of the metal to prepare it for finer tools, or those which, by working

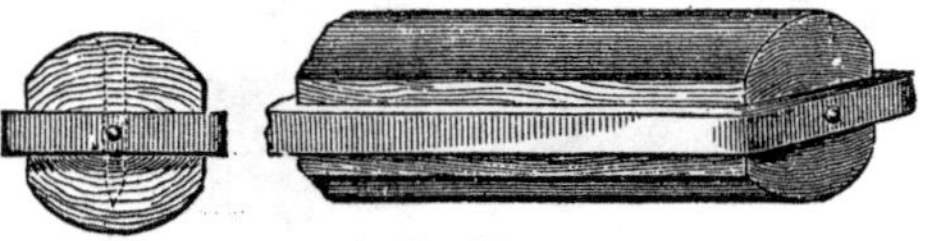

Fig. 30.

with lighter cuts and sharper edges, leave smoother surfaces. To produce a smooth surface in iron, sometimes a scraping tool, called a bit, is used, as shown in fig. 30, and in other cases the tool shown in fig. 27 is modified and shaped as in fig. 31; both amount to virtually the same thing. The bit is merely a flat steel bar with an iron shank. The edges of the bar are turned to the proper size and then filled up so as to clear behind. The extreme ends are slightly rounded and the size a little smaller for half an inch along the length of the bit, so that the tool will have a fair entrance in the work. The pieces of wood on the back are either beech or hickory, well seasoned, and fastened to each other by screws passing through holes in the steel. With a properly made bit the most beautiful holes, true and smooth as it is possible to conceive of, may be produced. There is no limit to the size of them, within reason. We have seen bits of twenty inches in diameter used in the largest workshop in the country with excellent results. The pieces of wood are intended to steady the cutter, as every mechanic knows, and it is not "just as good" to pack them up with paper or thin board when they get worn down, as they will after awhile. Two pieces of wood is all that should be used, for when packing is placed beneath it gets loose and shackly, and the cutter does not work as it should. In the

Fig. 31.

engraving the wood is rather short; it should come up to the end, or else there will be no support for the bit when first started.

This is not a new tool by any means. It is a very old and well-tried one, and while we advocate progress in every thing, we do not reject good tools because they happen to be old for the sake of new ones, simply because they are new. A substitute for this tool is found in the steel bar shown in fig. 30. It is called by some a pod auger, but is not capable of doing as fine work as the wooden-backed flat bit, while it is much more costly to make and repair. This is one form of a pod auger (fig. 32).

Fig. 32

The shape is sometimes varied slightly at the cutting end, *a*, but not enough to warrant a number of diagrams. The pod or body of this tool is a true circle, and the cutting part is merely a sharp, strong fleam, or steel edge, projecting forward so as to clear the front end of the pod and give a cut to it. This tool will take hold only after a recess has been made for it by a drill, if the work is not cored out; if it is cored out, the boring tool must be employed to bore out a portion of the hole, so that there will be a true circle for the bit to start in. It is necessary also to have a dog on the shank so that the tool cannot turn, or

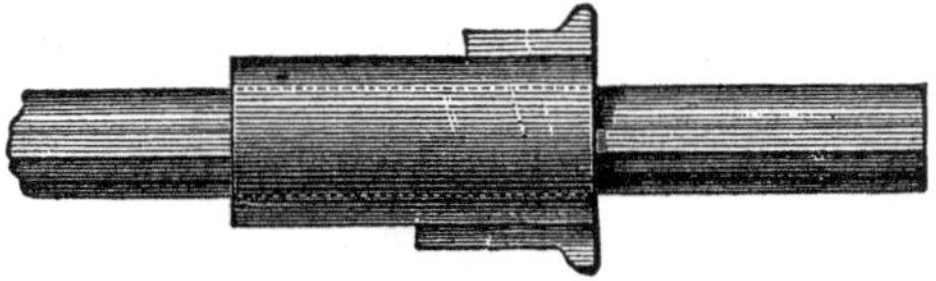

Fig. 33.

else to square the extreme end and put a tap wrench over it to effect the same object.

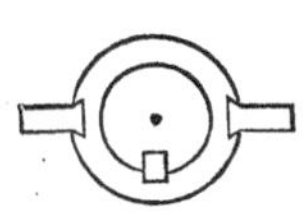

A much better, though more costly instrument, is illustrated in fig. 33. It is a very useful and complete one; and when properly made will do excellent work. For boring pulleys, or light, hollow work of any kind, such as small cylinders, valve seats, steam chests, cocks, etc., it has no equal. It cuts double as much as a pod auger can, and may (by increasing the number of the cutters to four) be made to carry a chip that no other tool will stand. The cutters must not all be set to the same diameter, but one inside of the other, so that while each will remove only a small chip, the aggregate amount of iron cut out will be as much as the belt will drive, or the chucks hold without slipping. The manner of using and making it is simply this: The mandrel is solid and made of steel; it has centres in each end, and also a key way or slot cut in it through the whole length.

The shell fits this mandrel nicely, and has dovetailed seats for the cutters, and a steel feather to fit the slot in the mandrel. There may be a number of cutters, and the last one may be set to take a second cut, so that the job will only require two cuts in all to finish it completely.

It is not possible to take a fine cut in connection with a heavy one, and have the work true; for no matter how well the shell may be fitted to the bar, the heavy cut will jar the light one so that the main object of it is defeated. We can take what we have called a "second cut," which means to allow the first roughing cutter with one set not quite so rank, so that when the shell has traversed the hole it will be nearly smooth, and certainly round and true. When the real finishing cut comes, the hole ought to be perfect.

The bar is to be set between the centres of a lathe, and the square end of a tool placed against the back part of the shell so as to force it along when the feed is thrown into gear on the slide-rest carriage. The hole need not be previously made true with a boring tool as it must with the pod auger.

Some lathes have a chronic indisposition to bore parallel holes. This is most annoying to those who like to have their work done well and quickly. The trouble is caused by the spindle not being parallel with the slides on the shears. A simple way of remedying the defect is to take up the back-head and put in a lining between it and the V of the slide it sets on. The lining must of course be so placed as to throw the spindle true, and the amount required will depend on the irregularity the lathe has. But with the "shell arbor," which we have just described, it makes no difference how much the lathe bores out of truth, for the truth of the bore depends altogether upon the position of the centres in the head-stocks of the machine—if these are exactly in line with the spindle (and they can easily be brought so), the tool under discussion will make a positively true hole, and if a tapering orifice is desired it is quite as readily produced by setting the tail stock of the lathe over to the desired point. It is difficult to conceive of any one tool more useful than this, or one capable of greater changes or applications compared with its first cost. It is really indispensable to every well-ordered machine-shop, and the intelligent mechanic will discover many nice points in the details of its construction, which we have omitted simply because we cannot devote too much time to one tool alone.

There are very many instances in the operations of the machine shop where special tools can be employed to great advantage. When large numbers of valve-seats and their chambers have to be made of an exact size, as it often happens in marine work, a tool which would make every one a fac-simile of the other, as regards the bevel of the seat, its distance from the outer flanges, and the diameter of the hole the lower stem of the valve works in, would be of great service. We present in fig. 34 a plan of such a tool. We can

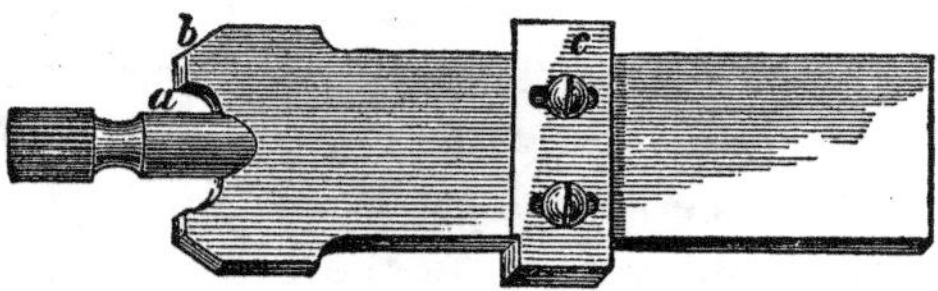

Fig. 34.

testify to its being a great economizer of time, as well as doing the work in a superior manner.

The work is roughed out first to nearly the size required by the drawing, and the hole for the valve spindle also bored. At the end of the bit there is a short rimmer, and immediately above it a solid shank, *a*, which fits the rimmed hole and steadies the bit while at work, so that the seat, which is cut by the edges, *b*, will be smooth, and free from chatters or irregularities. The cutter, *c*, is on the same line as the other cutting edges, being made to span the body of the bit in the position shown in the engraving.

In fig. 35 we have interposed an engraving of

Fig. 35.

another counterborer. It is merely a steel bar, having cutters forged upon it in the manner shown. There are an unequal number of these cutters, five being preferred, and after the tool is forged it is turned in the lathe and filed up so as to cut. This is a neat-looking tool, and one that we are assured does good work in the hands of skilful men. It may be made of any desired size or length; the one shown in the engraving is designed for gun work.

## CHAPTER VII.

### BORING TOOLS—CONTINUED. ABUSES OF CHUCKS.

THE tools shown in the engravings given previously are merely those which are employed in comparatively light work, and in the minor operations of general machine work. There are cases, however, where these tools are not available, and others, entirely different and distinct in character, must be produced. An instance of this may be found in the cranks of heavy marine engines. These are forged solid, and the holes for the shaft and crank-pin are cut out of the mass of metal. In old times

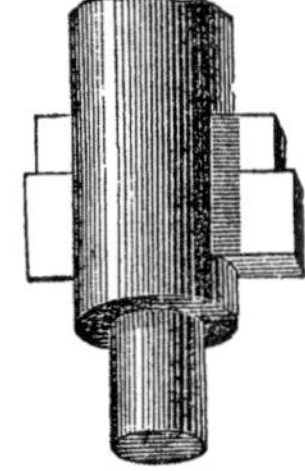

Fig. 36.

these cranks were bored by making a large hole with a common drill (say five inches in diameter), and afterward inserting a boring bar and cutter like the one shown in fig. 36, and enlarging the hole. This method is doubtless still practiced in many shops, but there is another way which is more expeditious and economical. This is to bore a solid core out of the boss of the crank, as shown in fig. 37, and leave the centre standing.

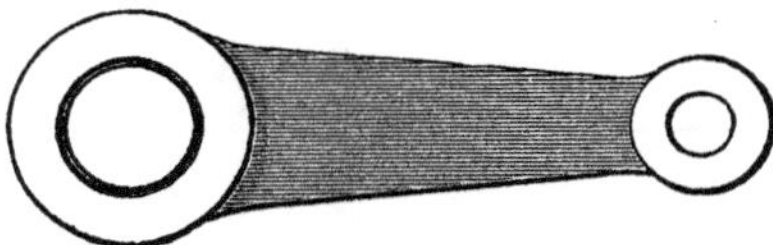

Fig. 37.

To cut out a hole twenty inches in diameter in solid metal is quite an achievement, and requires not only peculiar tools but careful superintendence during the operation. The tool is shown in fig. 38.

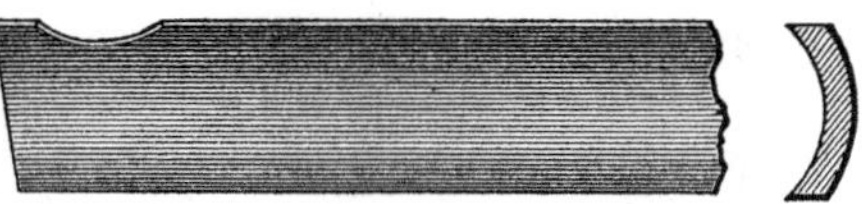

Fig. 38.

It is curved in section, and made very wide sideways, but not on its cutting edge. The cutting part is about ⅝ths of an inch across. Four of these tools are set in a cast-iron cross, which screws on the spindle of a heavy boring mill, and the tool thus arranged is shown complete in fig. 39.

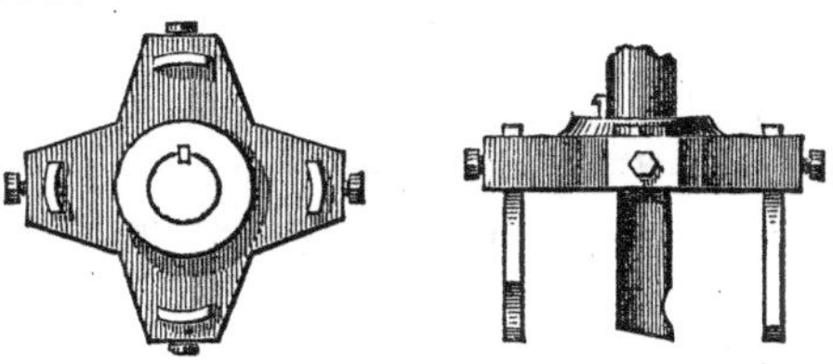

Fig. 39.

The cutters are set so that two travel in line with each other, while the other two cut in another track, so that in this way a wide groove is produced, from which the chips can be removed with facility. The channels are shown in fig. 40. The tools do not bind or clog when care is taken, and they are so wide that they do not spring side ways. When one side has been cut half way down, the crank is turned over and bored from the other until the two cuts meet. The central core then falls out. The hole is afterward bored true to the size required by an ordinary turning tool, and generally needs only two light cuts to finish it. With good luck one man should bore a twenty-inch hole, twenty inches deep, in fifteen or twenty hours. The economy of this plan, as compared to the old one, is striking, and should be practiced in all shops that do work of this class.

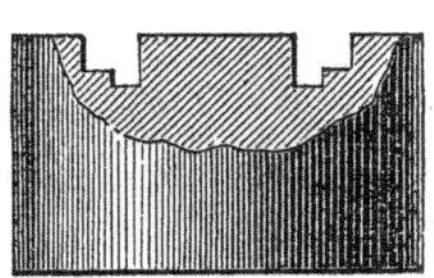

Fig. 40.

A plan for a boring bar and cutter, which was (and it may be still is) used in a mill for boring car wheels, is shown in fig. 41. These wheels are used in such numbers on long lines of road, that it is necessary to provide some means for boring them as fast and as economically as possible. With this cutter a car wheel $3\frac{3}{4}$ bore and 8 inches deep, cored out $3\frac{3}{8}$, has been bored in from six to eight minutes complete. The arrangement is merely an ordinary bar, with a cross-bit or cutter through it; but at right angles with this there are two steel rimmer-blades, dovetailed in the bar. These rimmers are turned up in the bar itself, and can be driven out for grinding or other adjustment as required. They taper very slightly from the bottom to top, and are made a little larger than the cutter, so as to follow it and true up the rough portions or surfaces left in the rapid descent. This cutter and bar does good work, when it is not forced too much, and we have known thirty wheels to be bored on the machine it was attached to in ten hours.

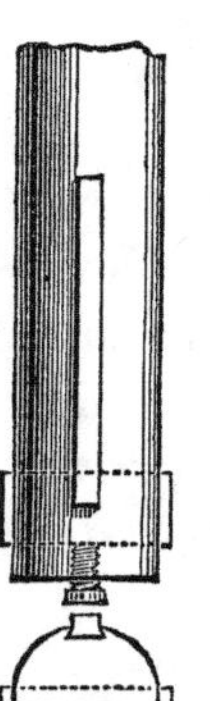

Fig. 41.

In the drills, and all other tools used by mechanics, there are innumerable cases where tools are made for special purposes, and it is principally for this reason that the subject is inexhaustible. An elaborate treatise on tools would present little that is really new; and to the practical reader there is no benefit in discussing those which have been used for years, unless some errors in their construction can be pointed out and removed, as in the case of the boring tool we illustrated in the earlier pages. For boring cylinders, and hollow work in general, where a bar and boring head is used, a cutter like the one shown in fig. 42 is very serviceable, but the

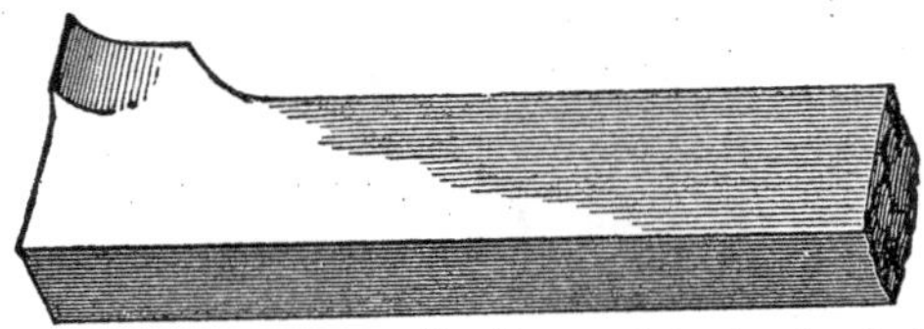

Fig. 42.

kind of work varies so much that one tool cannot be used continually, and the good sense and ingenuity of the workman must be the judge of what is required. Much also depends upon the feed and speed of the cutter, or the work, and unless these are well regulated, either the job is much longer in the lathe than it should be, or else it is not properly done. These details cannot be put down positively, for it very often happens that the intelligent workman does not know himself at what speed he will run until the job is under way.

It seems hardly necessary to exclaim here against the habit of idling over work that some individuals practice. "Slow speed and fine feed," say these gentry, make the job last longer; they are correct, undoubtedly, but they should also remember that the trick also makes their wages shorter. Men are paid for the work they do, and he that accomplishes the most and the best, will assuredly stand highest in the estimation of his employers. Let us all—as practical men—aim to drive the machines faster; have the cutters sharper, the feed as heavy as the job will bear. Let us make American engines and American machine work our pride and boast, and create a market for it all over the globe, and as a preliminary step to renown, criticize closely every thing that promises to improve the character of the tools we work with.

---

## ABUSES OF CHUCKS.

It sometimes occurs in the operations of a machine-shop, that the ordinary chucks fitted to lathes will not take in the work to be done, and resort is then had to wooden blocks bolted to the face-plate and turned out to any desired form. Sometimes these blocks are screwed on to the spindle itself, but in either case they cost time and money to make. It would seem, from the want of care and attention paid to these necessary appurtenances of a machine-shop, that they were considered useless except for temporary purposes and that the only disposition to be made of them is to leave them around on the floor, under the vice bench, or in any hole or corner that is unoccupied by any thing else. Some men find them useful to batter mandrels or arbors into work they are about to turn, to sit upon at noon-time, to build a fire with in the mornings; they find them convenient to punch sheet iron upon; in short, wooden chucks are abused in an infinite variety of ways which seem to us altogether wrong. Put them away in a safe place like any other tool. Assuming that the block will not run true after being shifted from the lathe, it can still be returned and employed again for work approximating in shape to the first job it received. The block out of which the chuck is made is always the best piece of wood to be had, and it is poor economy to cut up lumber to use, or rather abuse, in the manner set forth above. And in this connection it will not be amiss for us to protest against battering up the centres of shafts or mandrels by carelessness. No good workman needs to have any remonstrance addressed to him on this score; but bad ones are continually guilty of the practice referred to. If a workman wishes to damage his reputation in the eyes of all intelligent artisans, he will take a heavy hammer and blunderingly whack away on the delicate centre, that should be as carefully protected as the pupil of the human eye. Such a course only results in mischief; in a well-regulated shop it is soon found out, and the individual committing this outrage on common sense should be immediately dismissed from the shop.

---

## CHAPTER VIII.

BORING STEAM CYLINDERS AND HOLLOW WORK—EXPERIMENTS WITH TOOLS NEEDED—CONSERVATISM AMONG MECHANICS.

To be reliable and useful, a steam cylinder, or indeed any cylinder in which a piston works, must be mathematically correct as to its diameter from end to end. They are not always so; sometimes far from it. There are several reasons which may be assigned as the cause of the irregularities, and these are the manner in which the cylinder is bolted to the carriage (when bored in a lathe), the kind of tools used in cutting away the superfluous iron, the rate of speed at which the cutters travel, the shape of them, and the degree of temperature the casting acquires while being worked. For all of these troubles there are remedies.

It is the practice in the best shops to bore the cylinders upright; take out a heavy cut at first, and bring the interior of the cylinder by successive cuts (say two), to within the thirty-second part of an inch of the size required. The remaining portion is then removed in the last cut by a tool which is neither a round nose nor a diamond point, but a combination of the two; a moderate feed is given to this tool, and the boring head, or its equivalent, started on its journey. The theory is that the round-nosed tool, with fine feed, makes a dead smooth surface; this on first thought might appear desirable, but reflection will show that it is not so. Dead smooth surfaces in steam cylinders, do not wear so well at the outset as those slightly raised or ridged; and this may be accounted for by the larger surfaces exposed and the more intimate relations of the structure of the iron, or of the faces opposed to each other. Thus: cast-iron rings in cast-iron cylinders, are apt to cut when new, unless they are very loosely packed. With the round-nosed diamond-pointed tool, the objection is that the edge of it will wear away quicker, but the cut will be clearer and freer than the legitimate round-nosed tool; it will heat the cylinder less, and we think produce better results generally.

An engineer of much experience has told us that he preferred to have cylinders bored in this manner, to having such very smooth surfaces as are commonly used, and gave as a reason for his opinion that the cylinders were insured a better and more permanent finish than when glazed over at the foundry. No rude workman need take these remarks as an apology for clumsiness or want of skill; a cylinder bored in this way requires more careful attention than one bored with a round-edged cutter. Very many workmen resort to the use of blocks of hard wood in the boring heads to prevent chattering or jarring of the cutters; when this fault occurs it is a proof that either the bar is too weak, or else that the cut is too heavy; other things may cause it, but these are the chief. It is therefore better to dis pense with the pieces of wood, for the reason that they are liable to force the tools into the metal. When the blocks run over little chips, the wood is either torn out or else the cutter is driven into the cylinder; they also heat the cylinder, and, in short, are more fruitful of injury than of benefit.

Some shops, when boring cylinders, ship a cross in each end of the casting through which the boring bar is thrust; the weight of the cylinder hangs on the bar, and the rectitude of the bore depends on the rigidity of the bar, the correctness of its revolution, and the fit of it in the centres or crosses. It is needless to say that no cylinder can be bored true with such an apparatus as this; the interior will resemble the barrel of the Irishman's musket, which was made to shoot around corners. When the tool arrives at the bottom of the cylinder it will certainly force the casting hard down on the top of the bar, and when the tool arrives at the top, of course the reverse will prevail; the casting will be driven up toward the bottom of the bar. It is then apparent that the bore of the cylinder will be a true copy of the orifice in the cross, through and in which the boring bar works; as the weight of the cylinder tends downward, it will soon wear the cross oval, and the evils complained of will be observed. We have seen cylinders of twenty inches diameter and five feet stroke, bored out in this way, but hope never to see another one so finished. Let us add, in conclusion, that all tools and equipments, of whatever kind, used in boring cylinders, should be true and correct in shape; the bars should run absolutely true, and the cutters should be of that shape which experience has shown to be the best for the purpose. The work will be done better and more expeditiously when such practices are observed, than when the reverse obtains.

---

## EXPERIMENTS WITH TOOLS NEEDED.

Theory is one thing and practice another, and sometimes it happens, very awkwardly, that the experience of the workshop refuses to agree with the laws philosophers lay down.

There is another and a very important point in the economy of the workshop, which is the power required to drive tools. Let us know what is the best form for a roughing tool. Out of half a dozen turners but one will be found who has a tool that cuts at all, the rest merely grate or tickle the top of the metal, so that some few miserable raspings are taken off. That this is a manifest loss to the company or proprietor is evident, and proceeds solely from a want of knowledge of the right principles. To obtain the knowledge in question we must experiment, not guess, and we think that a series of trials with a view to ascertain the best form of edge for a roughing tool would not be time thrown away.

A good plan would be to take a small lathe and a train of gearing similar to those used for churn powers. Let a pulley be applied to this gearing, and a belt from it directly to the lathe. A weight suspended from the drum of the gearing would represent the power. Now let a tool be put in the slide rest and set to work with a stated feed, speed, and depth of cut. The time required to run one inch, or more, should be accurately noted, and the tool removed and replaced by another. This in turn should be carefully watched, and the result recorded. In this way the diamond-point, the round-nose, the side-tool, the "no kind of tool," would all find their appropriate places, and the results would show very satisfactorily, if the experiments were well conducted, how much power was required to cut one inch, with given feed, and speed, and depth of cut. Of course the same shaft should be used for all the tools to cut on. The conditions would not vary with larger cuts and heavier feed. Another point gained would be the knowledge of how much horse-power, expressed in foot-pounds by the fall of the weight in a given time, was required for a certain number of lathes of a known length of shears and swing. Roughing off work is the heaviest that is done on a lathe, if we except cutting screws of quick-pitches, and the expression would be the maximum power required for a machine shop.

Much other interesting and valuable information might be obtained which does not now occur to us, for instance the loss of time and money through working with dull tools, or those that were too soft, etc., and we hope that some enterprising foreman or manufacturer will think it worth while to institute these experiments.

---

## CONSERVATISM AMONG MECHANICS.

Tradition is a good thing in its way, but mere blind reliance upon it sometimes leads men astray. The teachings of the past, applied to the arts, form what is termed experience, and by recalling to mind exigencies where extraordinary means have been employed to overcome difficulties, men perform duties with more ease and certainty than if they had not such memory at their service. The reader may ask, "Suppose a man has not had extensive experience in some branches of his business, how shall he thus familiarize himself with them?" We answer, inform himself by taking advantage of every means within reach that lead to the desired end. Conversations with practical men; consultations with books or papers devoted to the specialty he wishes to become acquainted with; these have an important influence which cannot fail to be an advantage to the student.

The mechanical ideas of this age of the world lead men ever onward; that is to say, that every hour discloses some vital question on which the masses of mechanics are ignorant because they have never given attention to the subject; as, for instance, the most impenetrable armor; the most deadly gun, rifled or smooth bore; the best forms for the hulls of batteries and iron-clad ships; and countless other points which will suggest themselves to all. This is why we say the spirit of the age leads ever onward, and hence the necessity which exists for investigating the labors of those who have preceded us. Is it not palpable to every one that the individual who has a knowledge of three or four different processes of doing the same thing, is a far more valuable member of society than he who adheres obstinately to his old-time method in the firm conviction that *it* alone is worthy of attention? Most undoubtedly. Yet we go over workshops and see men at work with tools that the best authorities have discarded long ago as useless, and have superseded them by more efficient ones; we see lathes in use with narrow shears, small spindles, light screws; planers with chains instead of screws or racks, and pinions, chain-feed on the lathes aforesaid, and other exploded and thrown-aside devices that time has outstripped and supplanted by more efficient ones. These are the old school men, and they would succeed much better in business if they took advantage of the discoveries and theories reduced to practice by other men. Pull out the old-fashioned machines and replace them with others better capable of doing the work! They occupy room and waste time every day that ought to have been economized.

---

# CHAPTER IX.

## TURNING TOOLS.

THERE is no branch of the machinist's trade which is more interesting or important than that relating to the lathe and its management. Of two men working side by side with the same lathes, and on the same kind of work, the same feed and speed, one will do much more tnan the other. We see this exemplified on piece work. Here th earnings of the workman are exactly in proportion to his skill, and though his comrades may take every opportunity to discover the secret of his success, he still outstrips competitors.

This is owing in most cases to the tools the skilful man works with. The unreflecting workman cannot appreciate some small matter in the construction of a tool, and suffers accordingly. He will most probably be contented to work with a clumsy tool, like the one shown in fig. 43, instead

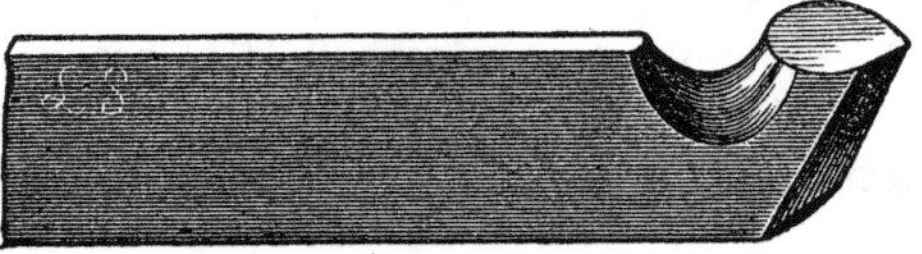

Fig. 43.

of the more efficient one illustrated in fig. 44, and he is perpetually wondering how it is that he is always behindhand.

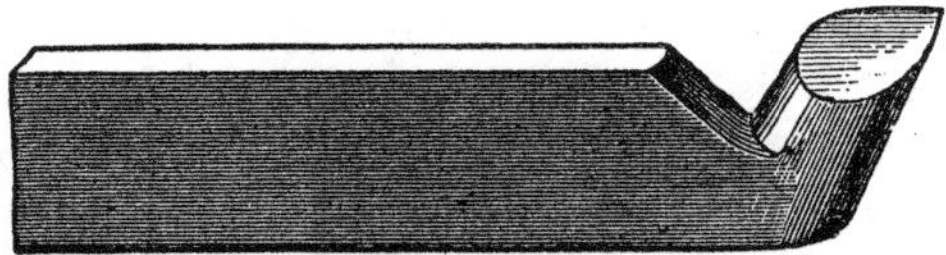

Fig. 44.

There is no mystery about the matter. A lathe tool works on one principle, as do all cutting instruments, and this principle is simply that of the wedge.

If a man has a heavy stone to raise, or a tough block of wood to split, he does not take a wedge which is thick and blunt, and almost as wide at the base as it is long. He uses instead a long, thin, and easy one, which does the work with facility and celerity. The case is exactly the same when we cut iron or metals of any kind. To sever the fibres or crystals we must have sharp, thin-edged tools, as thin as they can be made with economy. With these, and proper feed and speed, the work will be well done if intelligence superintend the operations. It is most essential that the tools be made sharp and kept so. If they are not, the work will be poorly executed. It is also of the first importance that the work be truly and properly centred. The centre is the point on which the accuracy of the whole job depends, and it will be apparent even to the unprofessional reader that it should be perfect.

Very many workmen are content to take a centre punch and make some sort of a cavity in the end of the rod, and "let it go at that," as the saying is. No good workman does this, but shiftless and indifferent ones do, and their work always shows badly compared with that done in a proper manner.

Every centre should be drilled. The drill need not be larger than the tenth part of one inch, in ordinary work, and the object of drilling is to keep the point of the centre in the lathe from bottoming. The centres in the work should be enlarged with a countersink, like the one shown in fig. 45. But when the shaft is too heavy to be used in this way a square centre is put in the place of the dead centre of the lathe, a dog put on the shaft, and the job set revolving. The back end of a tool is then put in the tool post and screwed up tight, and the tool brought in contact with the running shaft. If the work has been drilled properly, the sharp square corners make a countersink like the head of a screw, so that when the working centre of the lathe is put in the spindle it will have a fair, solid bearing in the job, as shown in fig. 46.

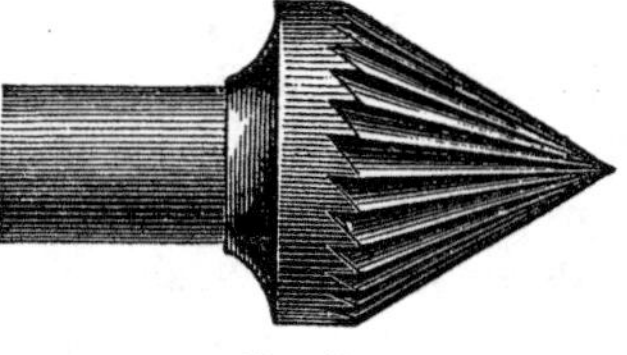

Fig. 45.

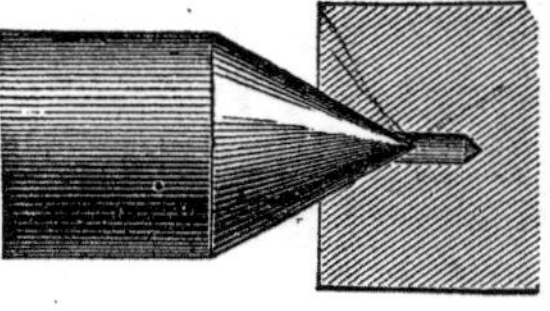

Fig. 46.

The way a centre, made with a centre punch alone, acts, is shown in fig. 47. Even if the punch is ground to an exact conformity with the lathe centre, which is by no means likely, the centre will not be true, as a rule, when the work is run over many times. For as the work revolves the

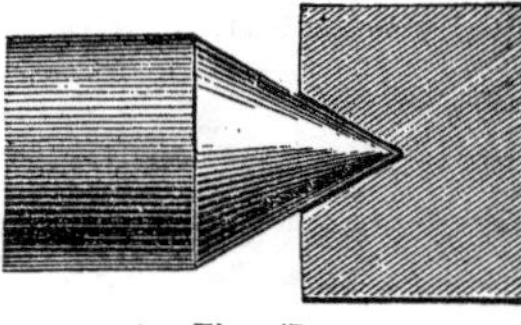

Fig. 47.

orifice in the end of the shaft wears, where it bears on the lathe centre. When the centre comes to the bottom of the cavity, as it soon will, it stops there because its point can go no further, while the larger or outer diameter of the centres wear away on the lathe centre. This causes the work to be untrue; when a rough cut is taken off from the shaft and a finishing cut is to follow, the work runs "out," and not only spoils the look of the job, by leaving rough marks in one side, but ruins the work, for it is not round, and can never be made to fit in its place. There are many ways of making countersinks for enlarging centres. One commonly used, quite as efficient, and much cheaper than the former one, is shown in fig. 48.

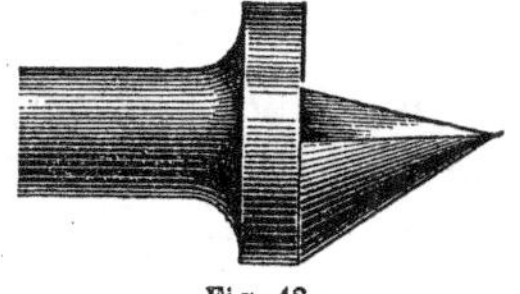
Fig. 48.

Having thus made a brief but necessary digression from the subject of turning tools, let us resume the consideration of them.

The tool shown in fig. 44 is a good roughing tool; it is called a diamond point, but there are very many turners who do not consider it the best for the purpose. It would be hard to say *why* precisely, for there is sometimes a great deal of whim exhibited in the matter of tools. Men will use, in spite of argument or reason, the tools they have been in the habit of employing, and prefer them to all others, even when they know they are not so good.

The cutter shown in figs. 49 and 50 is a most excellent one; its virtues have been well tried

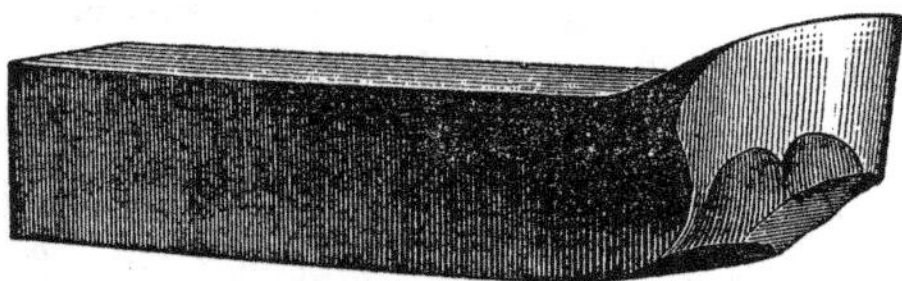

Fig. 49.

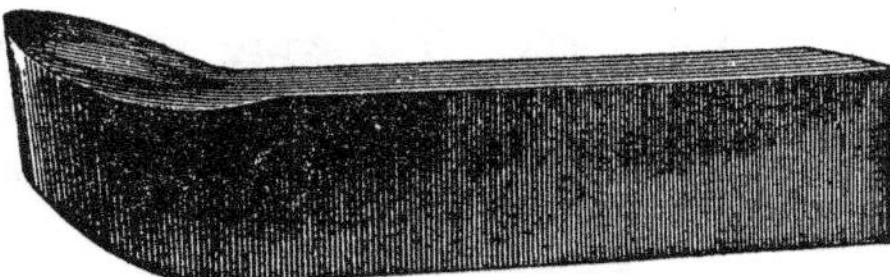

Fig. 50.

and not found wanting. It is stout, cuts well, when properly made, holds a good edge, and will carry a heavy or a light cut with equal facility. These are the chief requisites of a good roughing tool. The management of it depends on the workman.

## CHAPTER X.

### TURNING TOOLS—CONTINUED.

The best tools in the world, in the hands of a careless or indifferent person, are misapplied. Now, having obtained the proper tool, let us see what is required of it.

In roughing off a shaft of any considerable size there is hard work to be done; and, to economize power, every thing depends upon the shape and direction of the cutting edge. It will be conceded that to reduce the shaft with little labor to the lathe, and the least consumption of power, great surface in the cut must be avoided; that is, great surface considered in the direction of the length of the shaft. Of course, in reducing the diameter of a shaft a given amount, the depth of the cut is arbitrary and depends wholly upon the amount the work is reduced.

It is manifest that the round-nosed tool, here shown, which many mechanics use, is the worst that could be employed, for the surface of the cut

Fig. 51.

taken by it compared to that taken by the roughing tool already shown, is as the difference between a curve and a straight line. The round-nose tool takes a cut over one fourth of its diameter, and that portion of the surface is engaged; the shaft is not reduced in diameter, however, any more than by the diamond point tool, which cuts only on a small part of its edge, and works more directly to the end desired. The chips which are

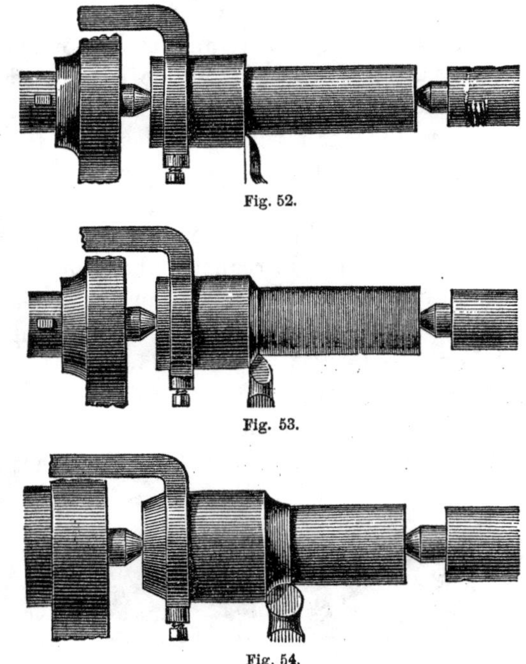

Fig. 52.

Fig. 53.

Fig. 54.

thrown off by these two tools show conclusively that one works harder than the other. One is long and free, and looks as if it were easily separated, while the other is compressed, short, and crooked, and its tensile strength much less.

These several figures present the idea clearly. The side tool, fig. 52, takes even less surface on the cut, in proportion to the iron removed, than a diamond point tool, fig. 53; but it is not an economical instrument to use for roughing off work, because it soon gets dull, and the point breaks off.

Let us now examine some kinds of tools for other purposes. After the shaft or the work whatever its nature, has been roughed out, it must in most cases be finished or polished. Polish is not always essential, but is indispensable in fine work, for a nice fit depends on the regularity or smoothness of the surfaces in contact. In former times it used to be the custom to turn a shaft as smooth as possible, to shift the belt on to the fast speed, and with a number of files, a great display of emery and oil, and polishing sticks, do what the turner should have done with the tool. The American mechanic of the present day knows a better process than this, which occupies but half the time. The emery used to get on the shear and in the feed screw, and wear them out rapidly; after polishing very large shafts, or other work, there was half a day's work to be done in cleaning the lathe so as to make it fit for use. The character of the finishing tool used in former times is but little changed, not enough to affect it, but the manner of using it is better understood. This is a good finishing tool, but success in using it depends on having it well tempered and ground,

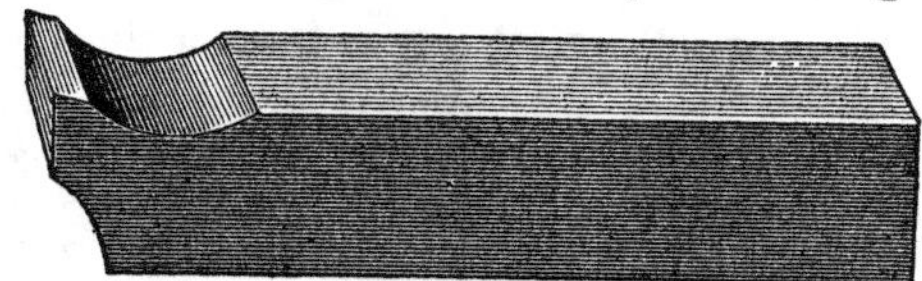

Fig. 55.

and carrying a light cut. Just enough should be left in roughing to take out the tool marks, and then the shaft will be true, and handsome in appearance.

This tool is not to be placed square across the shaft, or with its face bearing on it, but as in this diagram, so that the corner engages first and not

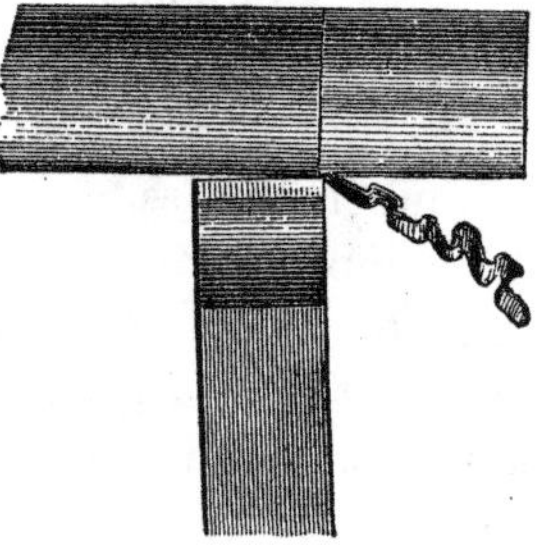

Fig. 56.

the face. In this position, if an irregularity is met on the shaft the corner is sharp and cuts it off, but where the face is engaged first, the edge glides over and the shaft springs, thus making an untrue spot.

This tool should in all cases be very slightly rounding on its face, in line with the shaft, and have but little clearance below, so that the cut will be nearly straight, for in this way it works smoother than when it rakes a good deal. Every tool has its use; the diamond point, or its substitute, for roughing; the round nose for fillets; the square nose for square corners; and so on to the end of the chapter; and it is as much folly to take one kind of tool for general use as it is to take one medicine for all ailments.

When a tool is properly made and ground it depends very much upon its position whether it cuts well or not. If the cutting point or edge, as the case may be, is set below the centre of the lathe, it will not work properly and is dangerous, for the tendency of the work is to roll up on it and leap out of the centres. When this occurs a double mishap is the result, for the work is not only injured, but the lathe shears and carriage are also endangered and oftentimes broken across, if the shaft be a heavy one. To cut well, the point of the tool should be slightly above the centre of the shaft, and the shape of the tool below should be such that no part of it bears against the job If it does bear, the shaft will not be true when done, and the tool will feed irregularly; sometimes it will jump in and take out a huge piece. at others it will not do any thing. This defect of a bearing on the work below the tool is shown in the appended engraving.

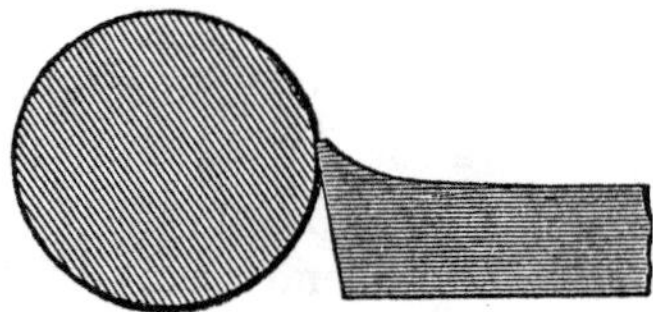

Fig. 57.

A side-cutting tool, or one that approximates to it, works in the same way when it has a bearing below the edge; the cutting part is crowded off by the pressure below until the spring of the tool forces the edge in heavily, then it takes out a "chunk" and stops until the feed forces it up again. Good, smooth, free, and true turning can only be produced by tools which cut where they should cut, and bear nowhere else.

## CHAPTER XI.

### TURNING TOOLS—CONTINUED.

A GREAT aid in turning long and heavy shafts and piston rods or shafting for pulleys, is found in what is called a "doctor." This tool is made in many different forms, but the essential principle of it is the same in every case. The object is to confine a number of cutters so that they cannot relax or slide away from the cut; for it will be apparent that should this occur the size would vary. The only change, therefore, in the size of the shaft will be from the wear of the cutting edges; where these are well got up, hardened, and set, the loss from such a cause is inappreciable, unless the work be full of sand seams. One kind of a doctor is made in the following manner:—

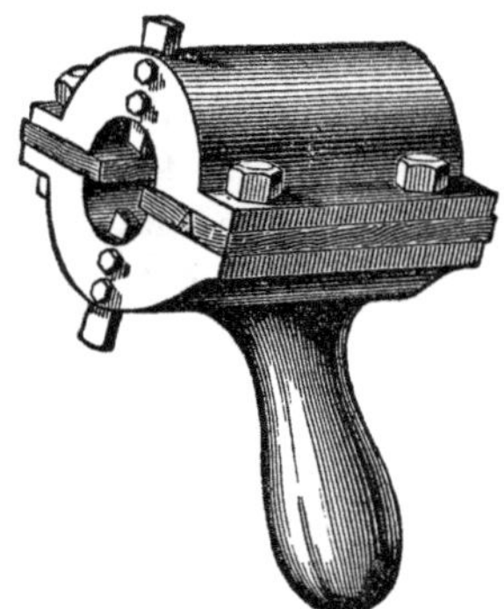

Fig. 58.

The outside casting is made in two pieces, and has a handle on one side. The wooden blocks, A, are placed between the two castings and screwed up tight. The distance between them should be just the size the shaft is when finished, and they serve to steady the tool so that the cutters will not chatter and make the work rough. The cutters are a little ahead of these blocks, so that the wood bears only on the turned part of the work. To accomplish this the end of the shaft to be turned must be started with a common finishing tool for a short distance, so that the blocks can be fitted to their places before the doctor is set to work. The cutters are ordinary finishing tools, ground square on their ends, and set so as to cut from their outside edges. The whole square face must not bear on the shaft or the tool will jump in. Very many mechanics round off the cutting corner of the tool, but this we think objectionable, for it takes more time to grind it, the tool has a greater surface to cut over, and the edge is more difficult to keep in good order than when square, or a very little rounded. The cutters themselves rest in slots, which they should fit accurately sideways. Backing them up with bits of tin or wedges is a slovenly practice, and takes more time than it does to plane the cutters so that they fit properly at the outset; they may be kept in place by set screws or clamps bolted over the end of the casting, and a tap with a hammer on the end will set the tool into the shaft for its cut.

In free, soft, well-forged wrought-iron, a doctor is a capital tool, but in scrap iron, in shafts that are full of sand seams and hard spots, it is difficult to make it work well. The sand takes the edge off the cutters and they have to be ground frequently. It is then a difficult matter to set them on the old cut, so that the shaft will be neither smaller nor larger than the part previously turned, nor yet have a shoulder or ridge where the cut was started anew.

Shafting for pulleys at the present day is all turned true and smooth. It used to be the practice to bore the pulleys a sixteenth of an inch larger than the shaft, or an eighth, for that matter, and put set screws in the hub so as to fasten the wheel in place. This way of doing work has been abolished. With such a plan the pulley never runs true. The belt is at one time slack and at another tight, so that the machine driven runs by jerks, instead of easily and smoothly. Grease and dirt also collect on the rough shaft, so that in time the upper part of a factory so fitted looks more like a hen-roost than the scene of organized labor.

It has always been a favorite idea with us to turn shafting in a lathe as broom handles are turned—that is, in concentric cutters. There is no reason why such a plan should not work with short lengths and small sides. In such a lathe there would be no tail stock or back centre, and the shaft, being carried in proper bearings, could be fed through the stationary cutters, just as a broom handle is turned in its lathe. The cutters might be set in a large box in the centre of the bed, and the cone pulleys on the lathe spindle, together with the head itself, should slide along, or the shaft might be fed through them. In brief, the shaft might be turned at the last cut just as a boring bar runs in its box. Let the cutters be in the box, and feed the shaft through; with water run upon it the finish would be beautiful, and the work true.

It may not be inappropriate to introduce here a fixture of the lathe which is often used in connection with long shafting. The common steady rest which accompanies a lathe is useless unless a bearing be turned for it to work on. With a shaft eighteen feet in length and two inches in diameter, this is a very troublesome and uncertain process, for the shaft is so long that it buckles and springs, and rides up on the tool, and often jumps out of the centres. It is to avert such a disaster that this fixture was contrived. It is a very old and very useful servant of the machinist, and is merely a cast-iron sleeve with steel set screws in it. This sleeve is turned true in the centre, so that it runs in the common rest, and is slipped over the shaft and secured by the screws. The turned part is

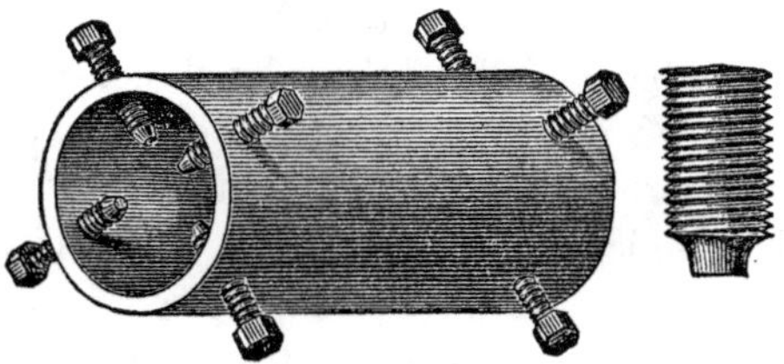

Fig. 59.

then set to run true—it makes no difference whether the shaft sags or not—and the steady rest applied. This is a short and simple operation and an extremely necessary one.

There is another tool, much used by some turners, which is called a spring tool, although its virtues are not apparent to us except for special purposes. It is made as shown below, in fig. 60, and it never "digs in," but goes about its work soberly and steadily.

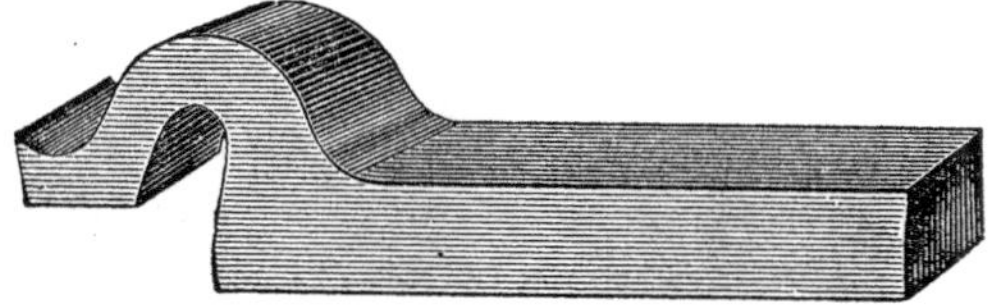

Fig. 60.

From its shape it will be seen that whenever an undue pressure is exerted on the cutting edge the same will give and recede, or spring down, which is the same, and thus prevent the edge from jumping in. For long shafts, or in places where the tool has to be extended a good way from the post, this will be found useful for finishing with, for it will not chatter if the surface in contact with the job, and the amount of cut, be small. We consider it of doubtful utility for finishing surfaces that require to be true, for every inequality that the edge comes to, if the spring part be strong enough, it cuts off, but if it be weak, it slips over and thus makes bad work. By putting a small wedge between the spring and the shank it can at any time be changed to a solid tool.

With a roughing tool and a finishing tool any one can turn out good work with a little experience, and observation will supply from day to day much more instruction than we could here impart. In complicated work, or in places where ordinary tools cannot be used, it may be of some benefit to our readers to bear in mind what follows.

The forked end of a connecting rod is a difficult thing to turn nicely. It is not troublesome to roughhew it, to make plunges at it with a round-nosed tool, to make chatters in it, or leave it in such a state that it will take a finisher three or four days to file it up. But to turn the various corners neatly, to leave the edges sharp, and the outline without ridges, is a nice piece of work, and on no other job can the turner show his ability better.

This is the piece of work spoken of, and although it is quite simple in its appearance, it is

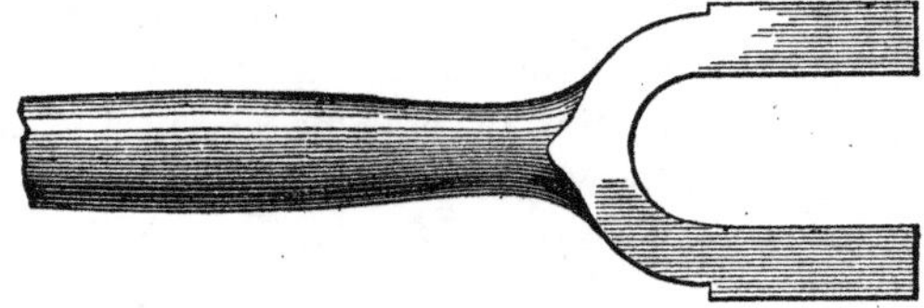

Fig. 61.

very troublesome. It is flat on the face toward

the reader, and unless the finishing and roughing tools are set at the proper angles, and well secured, they catch under the advancing edge and break off or jump in. Every mechanic knows what mortification it is to have a tool act thus; for when the surface has been finely finished elsewhere, one unlucky mischance by catching may spoil the whole.

As the rod comes from the forge it is rough, and in heavy rods for marine engines, such as we now speak of, especially so. If it is troublesome to turn the rod it is bad to forge it, and the blacksmiths generally leave an abundance of metal.

After the rod is laid out with the curves expressed on the drawing, and properly centred, the turner takes a square-nosed tool and runs in nearly to the lines all round, as in this diagram.

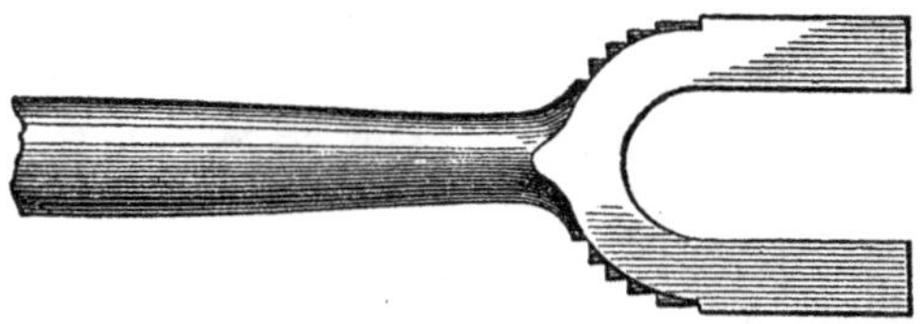

Fig. 62.

This roughs out to the outline neat and clean, and develops the shape perfectly. It is handier than any other method, because the workman knows exactly what he is doing. Instead of skipping about, taking off a lump here and a chip there, he goes steadily on to the end, and never makes one turn of the feed-screw handle without some advancement.

A square-nosed tool is better than any other for this purpose, because the edge, or corner, takes hold fairly and firmly, while the round nose, although it conforms to the curve better, is continually working or crowding off. When the tool has to be worked down a distance by hand, as in this diagram, it is better to put in an ordinary

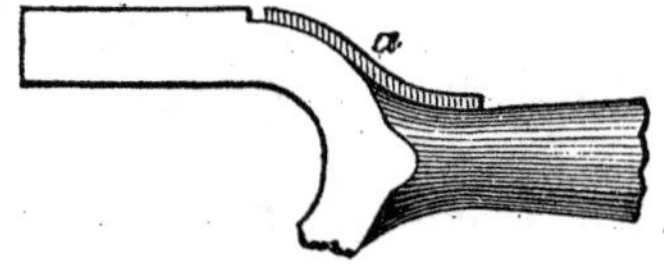


Fig. 63.

roughing tool, with the feed in, and start at *a*, and cut it right down at once to the centre punch marks denoting the outlines. In this way the lathe does much more work, for no man can feed as regularly and steadily, or as effectively, as the lathe itself can.

When the outline is once developed, and the ridges cut off by a bent side tool, the outline of the curves will present a surface consisting of a series of smooth-faced angles, without a rough cut, a "dig," or a chatter upon them. After this it is an easy thing to cut off the tops of these angles, and make one fair and beautiful sweep of the whole outline. The surface will shine as bright as the face of a mirror, and be as true as a pair of dividers can lay it out. We know this because we have tried it.

The final finish can be well given by a tool

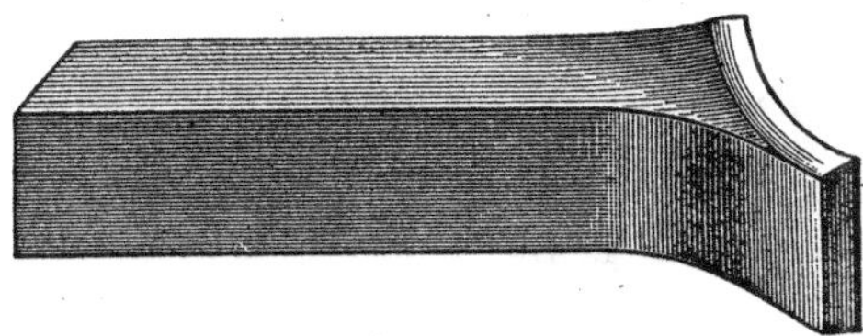

Fig. 64.

constructed as shown in fig. 64, and the reverse curve as in this cut (fig. 65). It must be borne

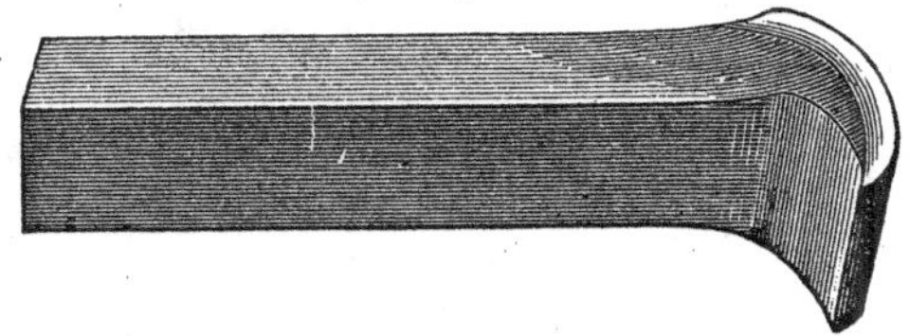

Fig. 65.

in mind that these tools have but little cut, or rake below, for the circle they cut on is very large and short, circumferentially, and a raking edge will jump in, while one too straight will push off. The linear length of the tool, or distance along the line of cut, should not be great, for the liability to spring is very greatly increased thereby. From two to three inches, and even less, ought to be sufficient for rods of ordinary marine beam engines.

## CHAPTER XII.

### TURNING TOOLS—CONTINUED.

VERY many mechanics start the carriage with feed down the lathe shears, and, as the curve falls in on the rod, screw the tool in, thus gradually working out the curve. This may be a good plan where the curve is large, but as it changes its character near the neck of the rod, and becomes concave, instead of convex, the handle must move over a long space very quickly, and it requires good guessing to tell just how much or how little the tool will take. Sometimes it misses entirely, or else takes a huge bite, and the latheman trembles lest the next thing his eyes behold will be the six ton rod flying from its centres and crashing into the bed below, while the general wreck of face plates, cone pulleys, etc., carried down by his mismanagement, tell a piteous tale of want of system and good workmanship.

It is not seldom that such cases occur. We once saw an impatient turner cutting a screw; when he came to throw the backing belt on to the pulley, he slammed the shipping bar with such violence as to spread the counter-shaft hangers overhead apart, and the shaft, belting, pulleys, and all, came thundering down within an inch of his head.

The ingenious turner can readily contrive tools for special purposes, so that by them his work will be greatly expedited. It sometimes occurs that jobs have to be turned inside and out at one time or without removal from the face plate or mandrel; ordinary tools are then inapplicable. Such an instance is shown below, where the casting has to be turned off inside and out without removal. As the inside cannot be bored with a boring tool (it being next the face plate), a special tool must be used, and one is shown in fig. 66, in connection with the casting.

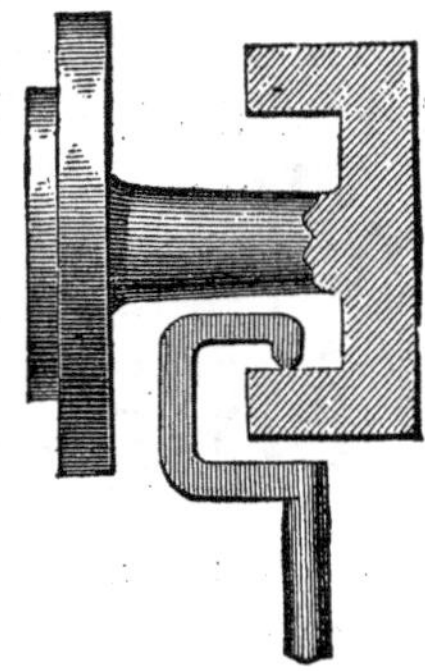

Fig. 66.

It must be borne in mind that we distinctly repudiate the use of such tools unless they are absolutely indispensable. The situation, however, is one that the turner has nothing to do with, and he must not be held responsible for want of good judgment on the part of the designer who contrived such an awkward piece of work. Such a tool as the one shown in fig. 66, springs and buckles because it has no direct support or bearing from the shank, and cannot be used at all with a heavy cut.

In the manipulation of heavy crank shafts much care and good judgment are requisite. Crank shafts for inside connected locomotives and screw engines are made in one mass, and it is a costly piece of work to finish them. For large marine engines, crank shafts of many tons in weight are sometimes built up or made in separate pieces and shrunk together. By this method they are not only as good as solid shafts, but better, for in the crank and pin the fibres of the iron run in the direction

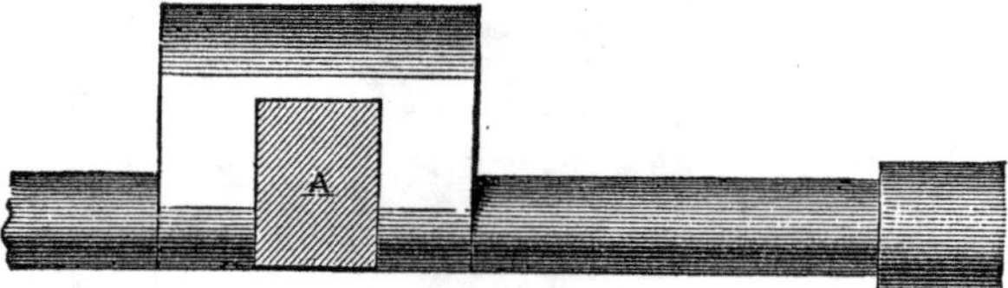


Fig. 67.

of the greatest strain. The case is quite different in solid forged cranks. When properly shrunk together the parts are immovable by any ordinary power. The *Golden Gate*, of the Pacific Mail Steamship Company's line, has a composite centre shaft of this kind, which weighs over twenty-five tons.

To return to the solid crank shaft (fig. 67). It is customary to cut the centre-piece, A, out at the slotting-machine, but this is sometimes impracticable, owing to the size of it, or other causes. It is also drilled out, but these several operations involve more labor than when done in a lathe. The block is first cut out by drilling holes along the line of the crank pin, as in this diagram, and

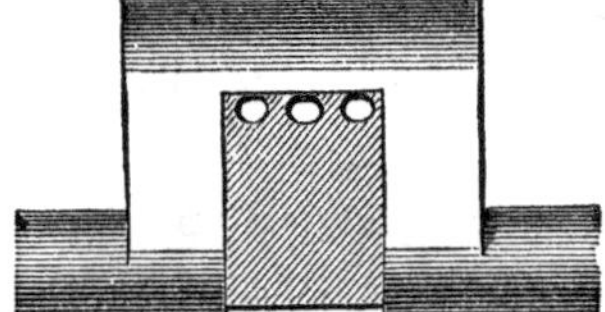

Fig. 68.

then running a square tool up to the holes. By this plan much handling of the shaft is prevented, for when the block almost drops out, the turner can detach it with a hammer and chisel and then go on and finish the pin up, without the trouble of taking out or putting it on the lathe so often. When the shaft is long it is a very troublesome piece of work to handle. There are strong cast-iron heads keyed on to the ends of the shaft, in which are centres to turn the pin on; balance weights must be put on the face plate of the lathe to compensate for the weight of the cranks. For want of these balances the work is very often spoiled. The crank being the heaviest has a tendency to fall forward immediately after passing over the top centre. The back lash of the back gear, which is always in, allows it considerable motion, so that a very little is enough to make it mount on the tool and break it off, or else cut into the surface of the pin and destroy its truth. All the centres must be well screwed up, and the lathe centres, especially, have a fair bearing, or else they will work out of correctness and make the crank pins and journals oval instead of round.

When the square-nosed tool is run in it must have a narrow steel shore under it, so fitted, in a slight depression on the lower side of the tool, that it cannot fall out when the tool springs, as it does after every cut; very many turners make the tool with a deep belly, so that it is strongest in the direction of the cut.

As the tool advances the shore advances with it, and the bottom of it rests in a shallow groove at the foot of the tool post. The tool should not have a lip on it, nor much rake, and the shaft must run slow and steadily. The feed must also be regular and even; and with these precautions there is little or no danger of jumping it into the work.

Another very difficult tool to manage is the common straight cutting-off tool. There is no reason why this should be so, but it is a fact and will be universally acknowledged by machinists.

The trouble arises from a want of care in making the tool. It merely requires to be straight from its cutting edge and sides down, not glanced or rounded off.

It is almost impossible to indicate in an engraving the slight amount of end roundness which will spoil the action of the cutting-off tool. If the corners of the sides are rounded over, even slightly, the tool is in danger of catching and breaking off, while, if the front or cutting edge be also an indirect line, it is liable to be drawn down instead of taking a direct hold on the work.

In connection with this subject we have already given an illustration of a straight finishing spring tool, and specified some of the uses cutters of this class were applicable to. They come in play in putting the final touches on the crank pin, for here the tool has to be extended a long distance from the support, and a common solid tool is in danger of springing and forcing the

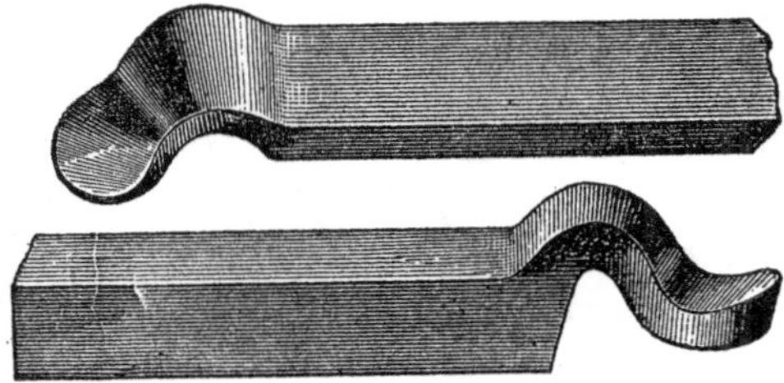

Fig. 69.

edge in. The spring partition must be made light enough to yield when a heavy strain comes on the edge, and yet sufficiently strong to carry a moderate feed without causing the surface to be irregular. The round corners are apt to be full of chatters when the solid or stiff tool is used, but with this tool the fillets will be, when used in connection with water, of the most beautiful character that it is possible to imagine. In these figures (69) an illustration of a round-nosed filleting tool is given, which works well and gives good satisfaction when properly used.

## CHAPTER XIII.

### TURNING TOOLS—CONCLUDED.

As grinding a tool and keeping the edge in proper condition is very essential to success, it will not be amiss to state a few facts of importance in regard to it. Inexperienced turners always go on the wrong side of the stone to grind; that is, when it runs from them. Every tool, no matter what its character, should be ground with the stone running toward the workman, as in fig. 70, the direction of motion being shown by the arrow. The reason for this is apparent to any one who thinks for a moment.

It is this—viewed through a magnifying glass the edge of every tool presents a serrated or saw-tooth appearance.

Fig. 70.

When the tool is ground with the stone running from the operator, all these fine threads, or filaments of steel, are drawn off toward the outside or upper edge, so that it forms what is known as a wire edge; the first application to the work breaks these off, and in a little while the tool is as dull as before it was ground. If, on the contrary, the tool be held against the face of the stone on the running side, as shown previously, the metal will be cut downward, and a keen, sharp edge produced, which will last much longer than when ground on the other side; it only requires an oil stone rubbed over it to remove the asperities and render the edge uniform. As the tool comes from the grindstone it is invariably rough, however smooth it may appear to the naked eye, and it is a good practice to touch up the edge preparatory to putting it in the tool post. It is this rubbing with the oil stone that gives that incomparable finish to wrought-iron when the tool is sharp. Such a polish is more durable than any that can be imparted with emery or oil, superior in appear ance and cheaper to produce; cardinal points in favor of using a sharp turning tool.

There are many tools which cannot be ground upon the stone without destroying the shape. Tools for forming beads or mouldings are of this class, but as they are generally used on cast-iron, they are intended to scrape rather than cut, and the faces can therefore be ground flat. It is generally easier to file the tool to the required shape and grind it when dull.

Tools that are filed have two disadvantages which make them inferior to those tempered and ground subsequently. When a tool is tempered, the smith dresses the edge by repeated blows, and compacts the metal at that point very closely, thus making it tougher and finer in grain. The hardening process is also an advantage, for the edge is less apt to be wiry than when the metal is fibrous; which is the case with annealed steel. A tool that is to be filed into shape must necessarily be soft previously, and though the workman may be an adept, he is very likely to slur the fine edge over in forming it, and make it rough and dull, instead of sharp. When the edge of a filed tool is tempered it is apt to crumble, and is, in many other respects, inferior to one that is ground.

For turning a moulding or bead on a side pipe, or cylinder head, such as the one shown in this figure, it will be found convenient to make the

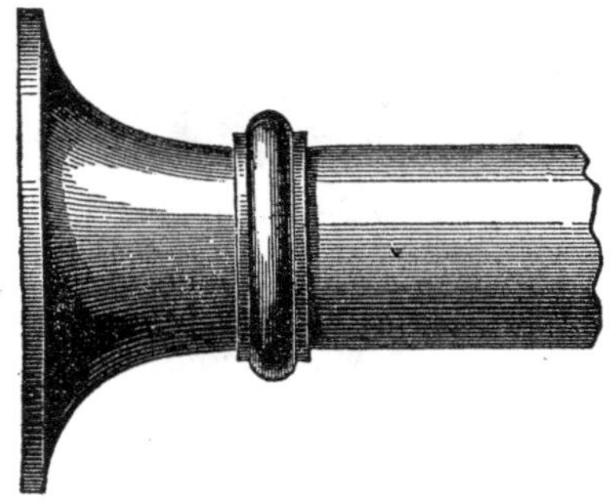

Fig. 71.

beading tool on the spring plan, illustrated in fig. 60. By this method it is less likely to chatter, or leave ridges, or cut roughly.

Of tools other than those used for cutting wrought and cast-iron, there are few which are materially different in external appearance. To this statement there is one exception. Brass cannot be cut by the same tools that are used for iron. Below, in fig. 72, we give examples of tools for turning brass. It will be seen that they are perfectly straight on the upper faces, and have no lips or acute edges. It is not possible to cut brass with a drill, or any other tool that has a cleaving edge. Such edges draw in to the metal and throw it out of the lathe, or else jam and break off. There are

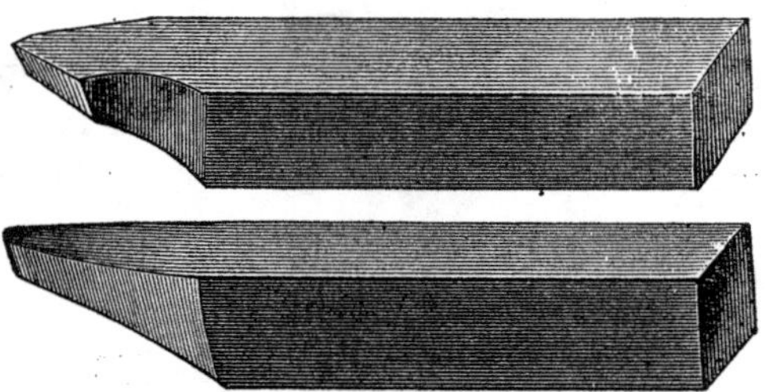

Fig. 72.

compositions of copper and tin, zinc and copper, and others, which can be cut by common tools, but these are not brass, which consists of specific portions of certain metals. One of these tools—the round nose—is used for light cuts, and the other where large amounts of metal have to be taken off at once.

In turning wrought-iron very many turners make their tools quite hard, and cut the metal dry or without water; preferring to absorb power rather than soil the lathe with sloppy combinations of iron and water. With proper care but little "muss" will be made, while the gain in time, by using water, is very apparent. Not less important is the power required to drive a given number of lathes. Those which run dry require more than tools used with water, for the simple reason that the friction is greater. Any one can test this to his entire satisfaction by putting a tool in a lathe, starting the cut, and driving the machine by hand. It will be found that when the chip is of such a size that the arm can hardly turn the lathe dry, the addition of water will free it immediately, and the lathe can be driven with ease. If the shears be well oiled previous to beginning a job, the water can be wiped off without injury to them, even though the work be days in progress.

This chapter concludes the series on this subject. The skilled turner will perceive many cases not laid down under this head which might have been alluded to, but it is impossible within the limits of our treatise to detail every minute manipulation a lathe is capable of. Special instruction on particular points has not been aimed at, but a general and familiar treatise on the tools used in turning.

# PART III.

# MISCELLANEOUS TOOLS AND PROCESSES.

## CHAPTER XIV.

### LEARN TO FORGE YOUR OWN TOOLS—MANUAL DEXTERITY—SPARE THE CENTRES.

MANY mechanics have an idea that after they have mastered the more legitimate duties of the workshop, they have learned all that is necessary and can undertake any thing in their line of business. Machinists particularly are prone to this error—a common one by the way—and think that a knowledge of fitting and turning, once acquired, makes up for all other deficiencies. In reality, the self-styled finished mechanic is, paradoxically, the unfinished one; for he who acknowledges his shortcomings, and tries to correct them by obtaining all the information he can, will acquire a more thorough knowledge of his profession. Comparatively few machinists are competent to dress their own tools, or, indeed, handle the blacksmith's hammer on any work. How many times such knowledge would have been invaluable, we leave individuals to decide from their own experience. A simple weld which they were unable to make, a faculty for dressing chisels without putting their own eyes in danger by striking the anvil instead of the tool, would assuredly have stood persons ignorant of such details in good service in time of need. Apprentices who go to the tool-dresser to have the edges of their chisels or other instruments renewed, will do well to observe the process and inform themselves of it. Observation and experience are twins and inseparable, and no youth, or indeed any adult, can hope to attain eminence or proficiency without paying some respect to the matter herein alluded to.

Nothing looks more slovenly or impairs the value of a tool quicker than the accumulation of dirt and grease in its joints or about its bearings. The filthy oil that most manufacturers use, from a mistaken idea of economy, forms a glutinous mass outside of the bearings of lathes and other machinery, in which cast and wrought-iron dust and grit collects, to the great detriment of the working parts. Aside from this fact, the drill shavings and chips from cutters, if allowed to gather in the bed, or about the foot of the tools in question, give the shop a slovenly appearance, which greatly prejudices it in the minds of observing people. A lathe or planing machine that is clean will do twice the work that a dirty one will, at less cost often; and over and over again we have watched some clumsy fellow wading around in chips, or else catching one up every now and then just before it fell into some of the gears. Such a man cannot do good work, because his mind is distracted by side issues.

---

## MANUAL DEXTERITY.

While the brain of mankind is invigorated and educated by correct study and discipline, the other parts of the body, more particularly the hand, and some organs, as the eye, can also be trained to tasks which at first thought seem wonderful and impossible. The Creator has so cunningly endowed our bodies that there is no labor to be done, no skill in artificing or fashioning the metals, that is beyond our reach. Even jugglers, who have no trade, depend upon digital swiftness, or the sleight of hand, to perform their "miracles" successfully; and the safety of rope-dancers depends not merely upon their balancing poles, but upon the degree of education they have imparted to their feet. If in such callings as these, wherein the sole object is to please the multitude, the culture of the members and organs of the body is essential to success, may we not say that in the mechanic arts, upon which such important issues now hang, manual dexterity is entirely indispensable? We would, therefore, earnestly impress upon our mechanics the importance of it. This, allied to intelligence, is what makes first-class workmen. It is by no means to be despised, for excellence in this respect is attended by many other qualities which are of the utmost service in the trades. It is an old saying that "the hand follows the eye;" this is only another form of expression for manual dexterity. We see the truth of it exemplified every day; even sportsmen shoot on the wing instinctively, after the first lesson of following the bird in its flight is acquired; and the machinist, when chipping iron, always hits his chisel on the head, even though his eyes be closed or his face turned from his work: this is manual dexterity. By tuition his hand has learned to work in that direction, and although in this case he is not guided by his vision in any respect, his blow is none the less sure. Let any one who desires to prove the correctness of this assertion, take a hammer and chisel, such as iron-workers use, and try to work with it; he will be speedily convinced that here at least manual dexterity is necessary to success and good workmanship.

Of two men working side by side on the same work, both actuated by right impulses, one will exceed the other just so far as he cultivates the motions and faculties, so to speak, of his fingers, all other things being equal. So much does the quality that we have made the caption of this article exercise its influence on men, almost insensibly, that we have seen artisans performing intricate tasks with an abandon and off-hand motion that was wonderful; and that, too, where the least false movement would spoil work which would cost them many a hard-earned dollar to replace.

---

### SPARE THE CENTRES.

The practice of knocking off the centres of turned work is a mischievous one. It is merely doing work that is not only needless, but that at some future day will have to be done over again. When a centre is once properly made in a shaft, or any other part, is is unalterable except by chipping or purposely changing its position; and work once turned true on good centres will always be true, provided no damage occurs to it. It is just in this particular that the true centre is useful, for if a shaft is bent, or an arm on one thrown out of line, the old centres are available and the injured piece can be made as good as new in a short time. Suppose, however, that the journal of a shaft is worn oval, or that the collar is battered and jammed up, how is it possible to find the true centre of the shaft? It never can be found; the shaft may be made to run straight, but not by its old centres if they have once been cut off. When shafts are forged too long, in cutting them to the right length great "tits" are left on the ends, which are both ungainly and in the way. This is the blacksmith's fault, and must be remedied by the machinist; cut the shaft to the right length first, knock off the centres if they are too long, and then re-centre the job and finish it according to the drawing. In steam-engine work especially, the centres of shafts are essential to nice adjust ment, and they should never be removed.

A foolish notion prevails among some mechanics that centres injure the finished appearance of the work, but it seems to us that this is an erroneous view which ought not to be tolerated. Drill every centre, and drill it deep; countersink it so that it will have a good bearing on the centres of the lathe, and the workman will have the satisfaction of knowing that, all other things being equal, he will have a good job, and one that can at any time be easily repaired if damaged.

## CHAPTER XV.

### ROUGH FORGINGS.

I HAVE often remarked, in the course of my professional experience, upon the indifference displayed in some of our large machine shops toward obtaining good iron forgings. In certain intricate shapes, where the safety of the work would be imperilled by too much elaboration, when often heated, where some heavy parts are in close proximity to some very light portions, it is perhaps advisable to bring the work something near the finished size and leave the rest to be removed by machines intended for such business. Instead, however, of working as closely to the drawing as they might, a great many blacksmiths leave altogether too much iron for the turner and planer to cut off. This practice is to be reprehended, as, in addition to the increased cost of the job, the value of it as material is very much reduced. If a blacksmith leaves from three fourths to an inch and a quarter of sound iron for the turner to remove from a shaft five inches in diameter, he is guilty of a very great waste of time, labor, and material. We do not allude to shafts turned up from rolled iron; any person who had to make a 5-inch shaft and should deliberately select a 6-inch bar of iron to turn it out of, would be regarded as demented by all sensible persons. If the practice is not to be tolerated in the case of rolled iron, how shall we reconcile the fact of forging a piece of shafting very much larger than there is any occasion for, with mechanical common sense?

Trip-hammers are very useful tools in a blacksmith's shop, for they condense metal into itself, and compact the fibres of it firmly together. What shall be said of those persons who leave such an excess of metal that the best of it is all turned off by the machinist at a dead loss to the proprietors? Comparatively a blacksmith can work faster than a machinist; he can heat his iron and dress off a piece of metal that would require four times the labor on the part of the mechanician. So also with heavy hammers; they can draw down an inch and a quarter of iron much sooner than a lathe can turn it off, and the shaft so hammered will be a far better one than another roughly forged.

In locomotive shops there are better forgings made than there are in the marine engine shops in New York. There is more die work and a greater attention given to producing smooth, sound, even, and good forgings than in the large works above mentioned. It seems to us that this subject ought to receive some attention. It is as easy to make a forging somewhere within range of the finished dimensions as it is to produce a lump of iron with scarcely the most remote resemblance to the final outline. This scale ought to be removed much oftener than it is. When iron is overheated the impurities in it work out to the surface; a certain portion of the exterior, a very thin skin of it, is burnt, this makes a hard, vitreous scale that ruins the edge of a tool in a short time. Every blacksmith knows very well how to knock it off and improve not only the looks of his own work but lessen materially the time demanded by subsequent operations. These matters are worthy of attention. They are those little details of machine work that are too often lost sight of, but which exercise a very material influence over the profit and loss account. A minute in a factory represents some portion of a dollar, whatever the same may be; it does not require any very brilliant effort of logic to see that many minutes make many fractions of a dollar. The waste of time in doing useless work has a pecuniary value, and it is just as foolish to cut an inch or half an inch off of a shaft, when it could be avoided, as it would be folly to throw money into the sea. Let us have no more such waste, but turn out blacksmith work in some degree approximating to the mechanical advancement of the age. We have seen shafts forged (and turned them too) that required to have two inches cut off the ends before they were of the right length. Such carelessness, for it is nothing else, shows a want of consideration for the employer's interest that should be seen to at once by those concerned.

# CHAPTER XVI.

## HOW TO USE CALLIPERS.

It may be safely assumed that comparatively few mechanics use callipers properly. There are many different forms of callipers, with which all mechanics are familiar, such as those having springs, and those which are secured, when set, by set-screws biting on an arc. While all these have their several merits, commend us after all to the old-fashioned sort, made with two legs, two washers, and a good rivet. Now what is the reason that one man will always make a good fit when turning a shaft to fit a bore, or the reverse, while another man makes a botch of it? The reason is that the former knows how to take a size, while the latter is ignorant of that duty. Sizes when turning are generally taken either with a pair of callipers or a standard gage. It would naturally be supposed that with the gage inaccuracy of measurement would be impossible. It is possible, however, and frequent, because the workman has not sufficient delicacy of touch to use the gage properly. Callipers are very sensitive, and are often used for extra nice work; if, however, the work has to be multiplied many times, then the use of callipers is not economical, and we must substitute some other method; moreover, gages are costly tools, and but few shops are able to own complete sets; for general work, therefore, we must rely upon the callipers. Blacksmiths have tools which resemble callipers, but they are uncouth and rude. The majority of men of that calling, when using them, set their callipers somewhere near the size they want, and then, upon comparing the work with them, *jab* them over the rod or shaft, as if they were going to cut it in two. The consequence is that the size of the work finished depends very greatly upon the resistance which the joint opposes to the blacksmith's strength. It is needless to remind the machinist that violence or pressure, applied to tools of this kind, only distorts the measurement and results in "bad jobs." The object with some workmen seems to be to find out how much the callipers will spring in going over a shaft without altering; not to ascertain how much metal must be removed before the requisite dimensions are attained. It is safer to go by the sense of sight than it is by the sense of feeling, in all cases where it is practicable to do so. When we can see that the callipers barely touch the object measured, we know it is of the proper size, but when we only feel of it, accuracy depends almost wholly upon a delicacy of feeling which all persons do not possess. When we say accuracy, we do not mean hap-hazard accuracy, that will admit of being somewhere in the neighborhood of the right size, but we mean absolute mechanical integrity, such as is obtained in the sewing machine and in the manufacture of our best steam-engines. Many new inventions are rendered useless and thrown aside as impracticable, solely because rudely made; let us then endeavor in the use of all tools, but more especially in the employment of those upon which the proper working of other parts depends upon good fits, to be as faithful as our abilities will allow us to be.

## CHAPTER XVII.

### A HANDY TOOL—RIMMERS.

HOLES in castings which are cored out very often, require to be made true and smooth, so that bolts will fit in them. Some machinists waste a great deal of time in plugging the holes up with wood and then drilling them out afterward; still others spoil rimmers and files in rimming or filing the sand out; it is needless to tell the intelligent workman that all these methods are costly and tedious. A better way to accomplish the object is to make a tool like the one shown in the accompanying engraving.

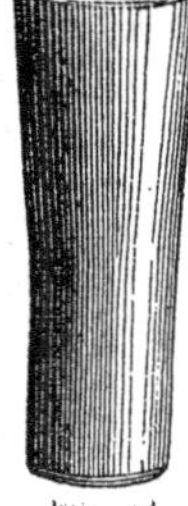

Fig. 73.

It can be made in twenty minutes, and is a simple but indispensable tool. It is merely a steel pin ground square on the face and turned true in the lathe. It may be parallel for a short distance (so that it will go straight) and taper above so that it will clear; the length is made to suit the work to be done. This pin is to be driven right through the casting, half from one side and half from the other, or else the face of the casting will be injured. With such a tool as this ten times more work can be done than with a drill or any other method, while the quality of it is excellent. It is called a drift pin, and may be made of any size.

---

## RIMMERS.

Rimmers are indispensable tools in all shops that profess to do good work. No matter how well holes may be drilled, they are not perfect unless rimmed. The twist drills now in use in the best shops make holes as perfect as drills can, yet even with them it is necessary to run a rimmer through where two parts are to be bolted fast—as a cylinder on its frame, a pillow block in its seat, or other details that require to be im movable.

The most common form of rimmer in use is the fluted one. The cutting part consists of many blades worked out of the solid metal either by planing or milling on a machine. These tools are good in many cases, but they are frequently made with too many and too sharp cutting edges. The hole formed by such a rimmer is not round, but a series of angles, as any one can see or feel by looking at it or putting a finger in. In our opinion it would be far better to make rimmers of this class with five or seven cutters than twelve or fourteen, as is generally done; and, furthermore, to leave less to rim in the work than is generally left, so that instead of taking a rank hold of the metal, the cutters would just clean the surface, and no more. In holes from half an inch to an inch, the sixty-fourth part of an inch is ample, if the drill is what it should be. In holes from an inch to two inches, a full sixty-fourth to one thirty-second should be allowed to rim out. Holes over two inches are cheaper bored out with a bar and cutter than rimmed, where it is possible, for the reason that rimming is done by hand, and is slow and hard work, while boring is done by power, and is quick and easy. Rimmers with seven blades require to be well backed off, as taps are, but not so much as to cause them to jam in the hole and work hard.

We have seen rimmers made with lozenge or diamond-shaped teeth, which worked very well. A pineapple forms a good natural illustration of their pattern. Such a rimmer is easily made in the lathe. To make it, put on screw gear to cut a quick pitch—say one turn in two inches for an inch rimmer. Cut a right hand thread and then cut a left hand thread on the same piece, plane it out, and back it off the same as any other rimmer. Such a tool will cut a round smooth hole and take more metal out with less labor than a straight fluted rimmer. Stubbs makes a five-sided, or pentagonal rimmer, with flat sides, that does well enough in a small work, but we never had a fancy for rimmers with flat sides. If it is necessary to straighten up a hole with a rimmer, and the tool is forced over to one side, a pentagonal rimmer is almost certain to bear in and work the hole oval.

Half round rimmers are very useful to boiler-makers or in rough work, but have no place in a machine shop.

A square rimmer is not worth a cent to do good work well. Holes, in castings that are cored out and have to be rimmed, should be drilled when over an inch, being first stopped with hard wood plugs driven in tight, so that the drill will have a bearing on the point. Holes up to and under one inch may be cleaned out with a drift pin, which is simply a square end punch. All rimmers, of whatever form, should enter the hole to be rimmed at least one inch before they begin to cut, so as to get a fair start and stand straight.

# CHAPTER XVIII.

## KEYING WHEELS ON SHAFTS.

WHEELS in machines are secured to shafts by different methods, such as forcing them on, by keys, by nuts, and washers, and occasionally by riveting; keys are more generally employed than any other device, being the surest way of preventing the wheel from turning, working off, or becom ing loose, when the respective key-ways are well made and the key properly fitted. We have been requested by various correspondents, at different times, to inform them where they could procure a work on key-ways, one which treated of the relative proportions for different sizes of wheels and shafts, and other general particulars which the experience of the author might suggest. We do not know of any, nor do we think such a work would be likely to meet with a ready sale; besides which, it would not be at all easy to lay down rules, or laws, for definite instructions in a case where so much depends upon practical knowledge. The principle of a key is that of a wedge, and it secures the wheel mainly by that force; how far the taper of the key should extend, and what material should be used for it, are matters which must depend wholly upon knowledge acquired by observation. In nearly all cases steel is preferable for small keys; but in some situations soft iron is better than the former, for the reason that it affiliates, or hugs, the shaft closer than a harder and finer-grained metal would. The sea-going steamers out of this port have large keys in their paddle shaft centres, by which they are held in place. It was at one time the practice to cast the eye, or hole of the centre, octagonal in shape; each plane of the octagon being truly filed to a bearing; the shafts also had corresponding planes, and when the keys were driven they were placed alternately with reference to the head and point; one key being driven from the right, the next one from the opposite side. In this way the centre was keyed up truly all round; the keys were large slabs of wrought-iron, from 14 to 20 inches long, by 4 inches wide, and 1 inch thick; the taper or "draught" allowed on these was not more than 1-16th of an inch; in some cases not so much. This plan of securing paddle-wheel centres has now been measurably done away with, as it was costly and not at all reliable, so many keys being used that one took the strain off the other, and some invariably worked loose. The method now adopted is to use one, or at most two, large and heavy keys. The centre is cored out in the foundery, so that only half of the circumference of the shaft hole has to be bored; this half is accurately bored to fit the turned boss on the shaft, and the keys are fitted to ways cut in the cored-out part of the centre; when they are driven, therefore, the bored part fits the shaft, and is forced into the closest contact with it. This plan is now generally pursued on all large steamers.

The hold of a key depends so much upon its fit and taper that, as we have remarked, individual experience must be the guide to success; but it is not amiss to assert that very little taper is necessary, and that beyond a certain amount the tendency of it is to split the wheel and cause it to work off. The wheel should fit nicely and then there will be still less strain required to retain it in place. In all cases gib-heads to the keys are a convenient means of drawing them out when such a course is necessary. We believe that car wheels are now pressed on, in the very best practice, and this will be found a good plan in most cases for other work. Besides being cheap, it is safe, although there is always a bursting strain on the hub which tends to weaken its endurance. Seats too deep and others too shallow in the shaft and wheel are to be avoided; the latter soon works off the corners where wheels are not well fitted, and the former makes unnecessary work and looks badly. In all cases the key must be proportioned to the work or duty the machinery is to sustain, and this proportion must be learned by observation, and an exercise of the laws of common sense.

# CHAPTER XIX.

## TAPS AND THEIR CONSTRUCTION—TAPPING HOLES.

A GOOD set of taps and dies is one of the most valuable properties in a machine shop, but the various forms adopted for them show that, sometimes, very little attention is given to the nature of the work required. The strain brought upon a screw thread is tremendous; in some places the lives of thousands of persons depend upon the fidelity with which the machinist has done his work; in any event, economy and good workmanship alike call for thoroughness. A discussion of the pitches proper for certain sizes of bolts is not necessary, as that question is pretty well settled now to the satisfaction of intelligent men; and if some unanimous action was held by those persons most interested on the question of adopting a standard, there would doubtless be very little further complaint made about uneven threads and fractional pitches. The office of a tap is to cut out certain parts of the iron and leave the others in relief; in plain terms, to form a thread by actually *cutting;* this is impossible with some taps, for by the angles of the edges cutting is impossible; bruising would be a more correct term

Some roughly-made taps are cut with a chaser and to complete the clumsy job are planed square on the sides. Such a tap is good for nothing but to raise a thread in soft metal, such as Babitt-metal, lead, and copper. It is not fit to use on steel or iron, because it does not cut its way, but squeezes the iron up into ridges. A thread of this kind has no strength, because the iron is crushed by the tap, and the fibres comprising it are twisted and torn by the passage of the tool. Taps are also made by cutting many grooves all around the circumference, which lead one way, like the teeth of a circular saw, only a little more rounded on the back. This is a good form for a tap that cuts in one direction only, or for a finishing or "plug" tap to run down, after a stouter one has formed the thread. The chief trouble with it is, that if the grooves are many in number, the edges of the thread or teeth break off and ruin the tool; this is certain to occur if the tap is turned backward; the threads will be shelled off like corn from a cob. Another form for a tap is to cut four grooves at equal distances up and down the body; these grooves are to be made with a round-nosed tool, and as the cut would be straight as the tool was fed down, the sides of the grooves must be run under, slightly, so that the teeth will be hooked, or hawk-billed to some extent; this form permits the tap to be used either way, backward or forward, without danger of breaking off the teeth or threads. Working a tap back and forth is an indispensable feature in tapping large holes, where the strength of the workman and the quality of the work render it improper to force the tap straight through. Of course, when the tap is large, the number of grooves must be increased, and for very small ones even less than four may answer.

All things considered we prefer this form of construction over any other. The object in making a tap is to obtain a tool that will do the work well, and be durable; these ends are attained in the plan mentioned. We have seen a number of "fancy" taps at various times, which would have answered for surgical operations, so keenly did they cut. Of this variety, one made like a half-round rimmer, or cut clear down to the centre, performed very well, except that it had this defect —it made the thread larger at the top than below, for it was impossible to steady it when first entered.

We have also remarked the mischievous practice of using chasers on taps; such a tool is not needed, and is obviously a damage instead of a benefit to the work in hand. Every tap should be finished in the lathe by the same tool that cut it, as it can be, by good workmen. No man can carry a chaser over a tap as steadily as a slide rest can move, and a little divergence of the chaser to one side or the other makes the thread uneven and irregular, or, as machinists call it, a "drunken thread" Tempering taps and dies has a very great effect upon the durability and execution of them; no matter how well the machinist performs his part, if the hardening is defective the time has been wasted. This subject will be discussed at some future period.

---

## TAPPING HOLES.

It is a fact, no less remarkable than true, that too little attention is given, in some machine shops, to the importance of tapping holes correctly and properly. Not only are the holes drilled too large, but the tap is allowed to take its own course, and if the bolt which is to follow in the threaded hole works as it should, it will be more on account of good luck than proper management. It was only the other day that we saw a workman upon an iron-clad, tugging away at a one-handed wrench and endeavoring to turn a tap that was beyond his strength. The tool was working badly and he was doing much more harm than benefit to the job, and we could not but reflect how much it might cost to repair a piece of recklessness which should never have occurred.

The consequences of abusing taps might be enlarged upon at great length, but we forbear, and content ourselves with simply remonstrating against threading holes out of all truth when they should be perfectly square—against drilling three-sided holes for a tap bolt—against drilling holes so small that the tap must be driven in with a hammer before it will "take"—against tapping holes in castings as they come from the foundery, full of scale (this we have repeatedly seen done)—and against the whole array of misuses to which these costly appurtenances of a machine shop are subjected. It takes time to make a tap, and as a great deal depends upon having them in good condition, more attention should be given to the proper use of them.

And while we are finding fault let us say a word about files.

---

## ABUSE OF FILES.

There are by far too many files wasted and misused in ordinary work, and the abuse is one that should be checked at once. To judge from the treatment some persons bestow on these costly tools, they are as common as pins and about as valuable. A new file is used for fitting a Babbitt metal box to a shaft, or a file for brass work is used alike on iron and brass; and then another must be procured when the workman desires to finish brass again. And so the interchange goes on, until the consequence is that the workman guilty of such carelessness has no file of any kind, fit for any purpose, in his drawer. Hard steel makes no difference to a file-abuser either. Apparently there are some individuals who think that because a diamond will cut another diamond so a file must bite another file; they pursue this theory and rasp away on the scale of cast-iron, or over the black places in forgings, with an utter disregard of their employers' time and money. A fifteen-inch flat bastard file costs from a dollar to a dollar and a half, but we have seen one of these tools placed *hors du combat* in five minutes by the blundering stupidity, not to say criminality, of the person using it. If the individual had been obliged to buy it himself, it is hardly to be supposed that he would have treated it in such a manner. It contributes in nowise to the reputation of any workman to be careless of tools that he uses but is not obliged to purchase, and it would be much better for all parties if a little more consideration were given to this matter.

## CHAPTER XX.

DEFECTIVE IRON CASTINGS — "BURNING" IRON CASTINGS—HOW TO SHRINK COLLARS ON A SHAFT.

IT is not uncommon to see large iron castings constructed with little or no attention to the expansion and contraction of the several parts.

Examples of the practice in question may be found in stationary engine frames. Cumbrous pillow blocks are cast upon them and immediately beneath is a large opening surrounded by sundry "filagree arms, scrolls, and similar articles," which, in the pride of his heart, the designer intended for ornaments. Still other instances of defective castings may be found. In turbine wheels, the step-frame, or that part which carries the weight of the wheel and shaft, frequently has large and heavy parts contiguous to light and thin ones. Large band wheels or pulleys are also examples, for from the solid hub and heavy rim spring light arms very much less in size and weight than the part to which they are attached. Car wheels of some patterns are open to the same charge, and many designs have been originated with a view to correct the fault. That it is not a trivial matter is shown by the results consequent upon malconstruction. Where the drivers of locomotives have cranks cast on them, the two arms which run to the eye of the crank are sure to break in a short time, and an outside connected locomotive can hardly be found that has not these two arms broken at the points designated. Even if the force of the steam were not exerted at that particular point, the jar and tremor when running would tend to disrupt the arms from the crank.

Iron bridges are sometimes made, whereof the girders and other parts under strain are cast with such a manifest inattention to the simple and well known law herein before alluded to, that the structure has given way and the public have condemned a system for the fault or ignorance of an individual. Many castings for different purposes, some to be employed in transporting passengers, some for purposes of commerce, are weak and fragile from the moment they are dragged out of the sand, because no regard has been given to a proper distribution of the strain of expansion and contraction. If these castings be struck with a hammer, the light parts will give out a clear high note, showing the tension to be great.

It is not only the breaking strain which is a consequence of bad proportion, but the difference in the quality of the iron composing the whole. Though the cupola may have been charged with metal of one kind the casting will not be alike when thick and thin parts are contiguous. Large masses of iron cool more slowly than small ones, the crystals are therefore coarser, and the metal less tenacious than small quantities of it, and it is therefore ill calculated to withstand torsion, compression, or tension, and many accidents that are apparently mysterious could no doubt be traced directly to defective distribution of the shrinkage.

## "BURNING" IRON CASTINGS.

The process known as "burning" iron castings together has long been practiced by mechanics. It often occurs that the too rapid cooling of one part of a casting causes an unequal shrinking of the mass, so that a tremendous strain is brought upon the weak parts. Corners of square surface-condensers, the inside angles of pillow blocks, cast in screw engine frames, the "gothic" arrangements sometimes perpetrated on the frames of land and marine engines, are liable to the contingency specified.

The loss of an entire casting from the cause mentioned, many hundred dollars in value, may be and has been prevented by "burning." The process consists merely in pouring melted iron on the fractured parts, placed in a mold or otherwise, as desired. When they attain the same heat as the liquid metal, fusion occurs at the points attacked, and the metal continues increasing in size until the operation is discontinued. Of course, a shapeless excrescence is formed outside, but this is readily trimmed off. Although not as sound as the body metal, it is still very strong.

We have seen hangers for shafting and spur-gear bearings mended in this way, and they afterward broke in an entirely new place, where the sound iron was.

## HOW TO SHRINK COLLARS ON A SHAFT.

In forging heavy shafts for marine engines or other work, the collars, where journals come, give a great deal of trouble. A very neat way of putting them on by shrinking is shown in the accompanying engravings, in which fig. 74 is a front

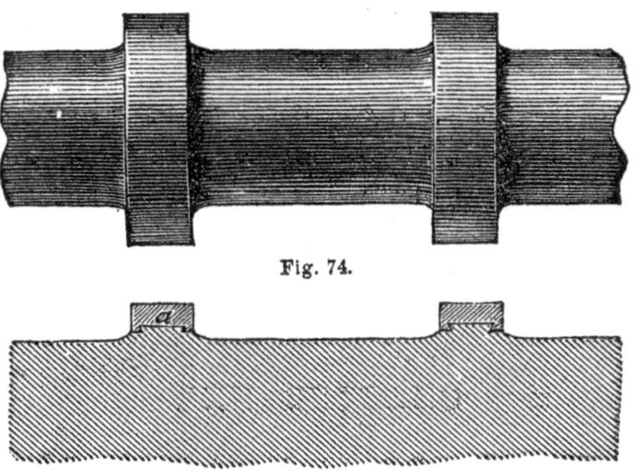


Fig. 74.

Fig. 75.

elevation, and fig. 75 a profile in section. The shaft is left slightly large where the journals occur, and bosses turned for the collars to set on. On these the seats for the collars are turned. The collars are welded up separately, and bored and turned complete in the lathe, a recess being left for the rib on the shaft. This rib need not be more than three sixteenths of an inch high in shafts twenty inches in diameter, being limited in size by the amount the collar will expand when heated. The lateral shrinkage left on the collar should not be more than the one hundredth of an inch; a snug fit will answer, since the collar never can shift even if it becomes loose.

After the several parts of the shaft are finished the collars are heated and shipped over the end of it, but care must be taken to ascertain first, whether the collar has been expanded sufficiently, otherwise it will stick when half on. It must not be heated so hot as to raise scale, for that would destroy the fit of the several parts. For heavy shafts this will be found an expeditious method.

## CHAPTER XXI.

### ARE SCRAPED SURFACES INDISPENSABLE?—OIL CUPS—DRILLING AND TURNING GLASS.

In stating this question as broadly as we have done, we disclaim at the outset any intention of dispensing utterly with scraped surfaces, or of erasing from the vocabulary of mechanical technicalities this detail of the workshop. The doubt has arisen in our mind whether much of the time and elaboration expended on scraping iron surfaces might not, without injury to the work itself, be omitted. The value of a positively correct face on a valve seat or on the V-shaped ribs of a slide lathe or planer, is undoubtedly great when it is well done, but when poorly executed the utility of it is, to say the least, questionable. We make the unqualified assertion that not one man in twenty is competent to finish a truly scraped surface. Scraping iron down to a perfect face is an art by itself, and comparatively little attention, so to speak, has been given to the subject in this country. The common method in use is to take an old file of any kind (except round or square), flatten its end out like a chisel, grind it up square on the stone, and then "grub" away on the iron wherever the workman sees fit. The chances are that previous experience has not fitted the operative for this branch of his business, and he mistakes a shade on the iron for a bearing and makes a depression still deeper by misapprehending the "situation." Of course the fallacy of attempting to make a true face in this way is manifest to every one familiar with the subject. It would have been far better to have saved the time wasted in such attempts and trust to good planing and attendance in future to rectify inaccuracies.

The better way to make a scraper is to form it like a Venetian stiletto, or, more familiarly, after the model of the section of a beech nut; that is, to have the blade triangular in section, and approaching concavity. With such an instrument, properly tempered, ground, and sharpened, the finest work can be produced. A flat-faced scraper is an abomination, and only fit to dig holes or to rough out the work for the triangular scraper; it is apt to make "chatters" in the surface; and when these occur we may bid a long farewell to any fine work without filing them out—a very pretty task to undertake after something like accuracy has been attained. Most scraped surfaces are nothing but a combination of scratches, shining blotches, and untruth; and while they are a waste of time to execute, they add nothing to the mechanical value of the work. We may fairly question whether valve-seats up to 180 square inches of area, say 15 inches by 12 inches, are benefited by scraping. In some locomotive shops in this country it is the practice to plane the valve-seat so that the tool-marks on it run in one direction, and place the valve so that similar marks in it cross the seat at right angles, and to set the valves running in this way without further adjustment. The results observed are that in a few days the valve has made a seat for itself that is far more durable than if it had been badly scraped. We do not go so far as some persons and assert that a scraped valve-seat is a positive injury, insomuch that the pores of the iron are filled with an impalpable dust that works out to the detriment of the engine in future. This theory is very finely drawn, although it may be partly sustained by facts. A finely-finished mirror-like surface on a valve-seat or lathe shears is indubitably of great value, and we must, in common justice, give credit to English workmen for great skill in this particular; in general they far excel our own workmen.

There is no reason whatever to interfere with the execution of a finely elaborated scraped surface in our own shops; but our observation convinces us that time spent in doing such work as we have seen, might be better employed in some other way.

---

## OIL CUPS.

A most objectionable and wasteful practice of using oil cans, instead of oil cups, for lubricating machines, prevails extensively. It is objectionable because uncleanly, for one reason, and extravagant because too much oil is put on at once. A journal will carry only a certain quantity of oil, and all that is poured in after the surfaces are well covered, runs off at the nearest aperture. When oil cups are applied, and properly used, the bearing takes up all the oil admitted, and uses it economically; that which is now lost might be saved. By an oil cup we do not mean a simple brass funnel to guide the nose of the can to the proper place, but a cup with a wick and a tube, or the equivalent of this device, for feeding the oil at regular and proper times. The wick and tube is the one generally used, and it can be made to feed fast or slow according to the amount of oil needed.

The filthy drip pans placed under the hangers of shafting are entirely unnecessary, and should be dispensed with by using cups. Many a suit of clothes has been spoiled, and not a little profanity caused by the upsetting of these drip pans, and the descent of their contents on workmen when belts run off. Where oil cups are not used, fully one half the oil poured on the bearing runs out again; and, as a matter of economy, every manufacturer, of whatever class, should see that his engines, his lathes, shafting, and similar machines and fixtures are furnished with oil cups that feed the lubricator to the journals, as fast or as slow as it is required.

---

## DRILLING AND TURNING GLASS.

Glass may be readily drilled by using a steel drill, hardened but not drawn at all, wet with spirits of turpentine. Run the drill fast and feed light. Grind the drill with a long point, and plenty of clearance, and no difficulty will be experienced. The operation will be more speedy if the turpentine be saturated with camphor gum. With a hard tool thus lubricated glass can be drilled with small holes, say up to three sixteenths, about as rapidly as cast steel. A breast or row drill may be used, care being taken to hold the stock steady, so as not to break the drill. To file glass, take a 12 inch mill file, single cut, and wet it with the above mentioned solution, turpentine saturated with camphor, and the work can be shaped as easily, and almost as fast as if the material were brass.

To turn glass in a lathe, put a file in the tool stock and wet with turpentine and camphor as before. To square up glass tubes, put them on a hard wood mandrel, made by driving an iron rod with centres through a block of cherry, chestnut, or soft maple, and use the flat of a single cut file in the tool post, wet as before. Run slow. Large holes may be rapidly cut by a tube-shaped steel tool, cut like a file on the angular surface, or with fine teeth after the manner of a rose-bit—great care being necessary, of course, to back up the glass fairly with lead plates or otherwise to prevent breakage from unequal pressure. This tool does not require an extremely fast motion. Lubricate as before. Neat jobs of boring and fitting in glass may be made by these simple means. The whole secret lies in good high steel, worked low, tempered high, and wet with turpentine standing on gum camphor.

## CHAPTER XXII.

### MANIPULATION OF METALS.

THERE are many occasions where a knowledge of some simple alloy or a peculiar solder would save hundreds, yes, thousands of dollars, just as a life may be saved by merely tying a pocket handkerchief tightly above a bleeding artery. It is only a few years ago that the valve-stem on the engine that runs the *Herald* presses broke in the dead of night, when but half the edition was run off. This was a dilemma, indeed, for a valve-stem is not made in half an hour, neither can it be bought at a hardware store like a pound of nails. The engine was injured in a vital part, and unless it was mended the entire edition would be stopped and incalculable loss sustained. Fortunately for the proprietors there was one of the employees present who understood the manipulation of metals, and he informed the bystanders that if they would collect their spare silver he would restore the broken part to a condition of usefulness.

It was done.

The stem was brazed with silver solder, and the engine performed until morning, so that the whole edition was successfully run off. But for the presence of the adept referred to, and his knowledge of this simple process, very great loss would have been incurred.

Some of our readers may be caught in just such a predicament, and we therefore append a formula for a solder which will braze steel. It is as follows: Silver 19 parts; copper 1 part; brass 2 parts; if practicable charcoal dust should be strewed over the melted metal in the crucible.

A good article of yellow brass is extremely desirable for the fine work in telescopes and optical instruments generally. A metal that works free and soft under the tool, and is capable of receiving a fair lustre from the burnisher, is always in request. A good yellow brass can be made from the following metals:—That denominated "watchmaker's brass" is made of one part copper and two parts zinc. German brass is equal parts of copper and zinc; the addition of a little lead makes the metal work easier and less liable to tear under the tool.

In all these mixtures the zinc must be added last, as it is a volatile metal and fuses at a much lower heat than the copper; the melting point of which is 4587 degrees, while that of zinc is only 700 degrees.

Iron and brass must be united by spelter, which is equal parts of brass and zinc. When the joints are cleaned and wired together fine powdered borax is applied to them as a flux. The solder is then dusted on in the form of a powder, or fine filings, and melted in, either with a blow-pipe or by being placed in a charcoal fire. Care must be taken not to melt the brass to be brazed. The solder of course has a much lower fusion point than the metals to be joined, else they would both run at the same time.

Aluminum bronze is a most excellent composition for boxes or bearings that run at a high speed, such as saw mandrels, fan blowers, etc. There is a small mandrel in Carhart & Needham's melodeon factory, New York, which runs 7000 revolutions per minute; it has aluminum bronze boxes, which are perfectly cold to the touch. Mr. Carhart informed us that he had tried every thing before this without success.

Aluminum bronze is made from copper, 90 parts; aluminum, 10 parts, and can be obtained in New York. Propeller shafts and boxes troubled with chronic heating might be cured by this metal. Boxes for fan blowers particularly, the shafts of which run from 3500 to 4500 revolutions per minute, might be easily lined with this metal. It is pronounced by those who have used it to be a superior composition for all journals at great velocities. Persons who are unaware of its merits will be benefited by remembering these facts.

A simple method of case-hardening small cast-iron work is to make a mixture of equal parts of pulverized prussiate of potash, saltpetre, and sal ammoniac. The articles must be heated to a dull red, then rolled in this powder, and afterward plunged into a bath of four ounces of sal ammoniac and two ounces of the prussiate of potash dissolved in a gallon of water.

These simple rules are practical, and will give good results with good workmanship. If the cast-iron is overheated and burned, the unskilful workman must not blame the formula for his failure; or if he put on such a blast as to blow the solder out of the joints, when brazing, and instead of making a joint spoils the job, he must not charge it upon us, but keep a brighter look-out in future. Good rules are useless unless put in force and practiced with skill and intelligence.

# PART IV.

# STEAM AND THE STEAM-ENGINE.

## CHAPTER XXIII.

### THE SCIENCE OF STEAM ENGINEERING.

NOT many months since a law was passed, requiring all persons in charge of steam boilers to have a certificate of competency from commissioners appointed to decide upon their fitness for their situations. In most instances, probably, this law has been complied with; in some others it has been wholly disregarded; in precisely how many we have no means of ascertaining; accident, however, has revealed one case at least, where the person employed as an engineer had no legal proof of his capacity, and, as the issue proved, no mechanical fitness either; he blew himself up with half a dozen others.

We shall not make unfounded charges, or be entangled in any assertions which we cannot prove; and we say that, although this is the only case that we know of, as being directly contrary to the provisions of the law for such cases made and provided, we can refer to countless instances where men have received testimonials of efficiency for engineering qualities which they did not possess; the five dollars of their employer bought them a character at second hand. Engineering ought to be a profession. It is not comprised in opening and shutting valves, in scientific flirts with an oil can, or in impertinence and vulgarity of demeanor when asked a civil question by an "outsider." It requires the closest attention, and both mental and physical labor, in order that the best possible results may be obtained; whoever does less than to devote all his energies to his profession, robs his employer and cheats the world of science of discoveries which he might have made had he used the faculties nature gave him. Admitting engineering to be a science and not a handicraft, we must then look for a high class of men to fill the situations—posts of honor and trust—which it opens out to the trade at large. No calling can be more productive of good results in respect to mental training than the one under discussion.

Familiarity with steam machinery, most especially with the boilers, is apt to beget a confidence in the ignorant, which is not born of a knowledge of the dangers and exigencies which are continually occurring during their working, and which is the offspring of conceit and the grossest folly; but contact with steam, a thorough elementary knowledge of its constituents, theory of action and production, only inclines the philosopher and the seeker after knowledge to be more patient and lowly in spirit when developing the mysteries of its sublime power, and applying the same to the arduous and monotonous task of doing the work of the world. A moment's reflection will show this to be the true light in which to view this matter, for in what other branch of the arts and sciences can we find another person into whose sole charge is given so much responsibility and power? The magazine he guards may spread havoc and ruin about if he impede the action of the feed, or neglect the valves which control the surplus pressure. If he be upon the railroad or in the crowded city, the weal or woe of multitudes is committed to his keeping; if he be upon the sea, in the shock of battle, when iron-clad answers to iron-clad, and the sea frets itself hoarse in the vain effort to overwhelm them, the fate of nations even is in his power; the cause of truth, of justice, and human rights, or the reverse of all these, lies hidden in the lifting of a valve, the lubricating of a rod or shaft, or the loosening of a gland or screw at the proper time.

These are not mere rhetorical assertions, they are living truths, every practical man knows it, and will give his testimony to the same effect; nor is it our purpose to especially glorify engineering above all other professions, but simply to direct attention to their peculiar sphere and duties, and to make them more conscientious in the discharge of their heavy responsibilities. Into their hands are given the wealth and property of the employer; not of one, but of many. Conceive, then, the delay and hindrance caused by neglect or mismanagement. Let one man in a district of ten square miles be five minutes late in starting his machinery in the morning; and reflect if there be one in such a predicament within the limit prescribed, what a loss will ensue to the country at large, through all its towns and cities. Or if this be too impractical in its bearings, suppose one careless man in the same area to squander his employer's oil, his tallow, waste, and small stores of all kinds; that man just as much robs his fellow craftsmen of wages as if he put his hand directly into their pockets, for the next one who comes after him will probably receive less to make up for the former's waste; thus, little by little, a trade or calling degenerates, until from being a profession or a science, it falls into the hands of incompetents and inexperts, and ceases to be any thing more than a mere occupation.

Only by slow, and sometimes painful degrees, can we arrive at logical deductions; and the study of steam engineering—one of the noblest sciences that ever attracted the attention of man—affords an example of the truth of this assertion. Among the mightiest physical forces of the globe, steam knows no master but a watchful one; it acknowledges no attentions but those which are undivided and unalienated by any other pursuit. Many employers, in their ignorance of the qualities required in an engineer, cause him to devote the intervals of firing to some other branch of their business, to saw wood, and, in one instance that we recall, to attend to the care of a horse; having, perhaps, some faint idea that this was the best way to give the man a proper conception of horse-power. This is not levity, it is not by any means funny; but only painful, as showing the estimation a noble servant of man is held in by those whose money is able to purchase its aid.

The relations of steam engineering and commerce are fully ascertained; it is not fitful in its action, neither spasmodic nor uncertain, but give its machinery undivided care and attention, and day after day it will go on its round of duty without cessation. The pressure will be evolved, the pistons rise and fall monotonously, and the whole grand and vast system of steam in this country will perform its functions without other derangement than such as usually falls to the lot of man's devices. It is only through attention to the subject of this article that superiority in it is reached. The expert attains to better results than the neophyte, yet the former was once awkward and rude in the science, and has only obtained his superior skill by a conscientious discharge of the responsibilities given into his keeping. If, therefore, each and every one in any way connected with the care of steam machinery resolves to raise the standard of his profession, the results will be apparent in a few years, in increased pecuniary benefit to himself, and also to the arts and world of science generally.

---

## CHAPTER XXIV.

### PISTON SPEEDS OF BEAM-ENGINES.

At one period of the science of steam engineering it was the practice to fix the limit of the speed of the piston at so many feet per minute; and from this and the other data usually taken into account—as the area of the piston, pressure of steam, etc.—the horse-power of the engine was calculated. If we are not in error, 250 feet has been set down as a standard speed for pistons: but modern engineers prefer to drive their pistons as fast as they can with safety, and to disregard rules which experience proves the uselessness of. We have, as a result, the performance of the engine of the *Golden City*, a new steamer belonging to the Pacific Mail Steamship Company. It is of the beam variety; the beam weighing upward of eighteen tons. This engine has a piston 105 inches in diameter by 12 feet stroke, and upon a recent engineers' trial trip, achieved the remarkable speed of 420 feet, or 17½ double strokes per minute. We have no doubt that the engine will be able to add materially to this speed, as the machinery was entirely new, it being merely an experimental trip. This is not an isolated case, by any means. The *City of Buffalo*, formerly a passenger steamer upon Lake Erie, now dismantled for the want of trade, had an engine with a cylinder of 76 inches diameter and 12 feet stroke, which drove paddle-wheels 34 feet in diameter, whose floats had 31 inches face, were 11 feet long, and had from 36 to 40 inches dip—19½ revolutions, or 39 single strokes per minute. By a severe exercise of mathematical knowledge, we ascertain this to be a piston speed of 468 feet per minute. We remember these facts and figures very well, as at that time we were pretty much occupied in looking after the engine aforesaid. The beam weighed nearly sixteen tons, and was stopped and started thirty-nine times in a minute, working with great ease and certainty. The beam of a beam-engine appears to some to be an insuperable obstacle to the general adoption of the class of engines to which it belongs; and its weight, momentum, velocity, etc., are charged heavily to its demerit. These theories, we fancy, are disturbed by the actual facts in the case, which are, that the beam is so poised and balanced on its centre that the supposed shock of changing its line of motion is utterly neutralized; and as for the weight—that is supported by the framing, and is no more against the power exerted by the piston than the smoke-stack. A beam weighing fifteen tons, or eighteen tons, can be moved through any portion of its arc of vibration, by the strength of a man; providing, of course, that the binders of the pillow blocks are not screwed up, and that the journals set fairly on the brass. The above cited cases of the speed of beam-engine pistons are all distanced by the extraordinary performance of the *C. Vanderbilt*, a Sound steamer, in her race, June, 1847. This engine is of 65 inches cylinder and 12 feet stroke, and on the occasion mentioned attained to 540 feet, or 22½ double strokes per minute. It is not at all uncommon or extraordinary to obtain a piston speed in beam-engines, of 400 feet per minute, in this country; but the performance of the *Golden City*, we think, is the best on record, considering the size of the cylinder.

Since writing the above, we have ascertained that all the facts just mentioned are below the mark. The *Mississippi*, a large paddle steamer, having an 81 inch cylinder and 12 feet stroke, has made 24 revolutions per minute, the wheels having 36 inches dip, and attaining a piston speed of 576 feet per minute. The *Metropolis*, a large Sound steamer, having a cylinder of 105 inches diameter and 12 feet stroke, has made 20 revolutions per minute, and we think a higher number. The working beam on the *Mississippi* weighs 14 tons; that on the *Metropolis* about 16 tons. The engine of the *New World*—a side-wheel steamboat 420 feet long, on the Hudson river, having a 76 inch cylinder, and 15 feet stroke, has made 20 revolutions per minute, or 40 single strokes. The *Richard Stockton*, however, has outstripped the whole fleet, and, we think, attained the highest piston speed for an engine of this class ever made in the world. We do not know the exact dimensions of the cylinder, but have been told it is between 50 and 60 inches, with 10 feet stroke. The *Stockton* has feathering wheels, and makes 32 revolutions, or 64 single strokes per minute; and has done this duty for years, having been built by Robert L. Stephens for the express object of testing the speed at which a piston could safely travel. This is the highest speed within our knowledge ever attained by a piston in an engine of similar size; if any other instances come to mind we shall place them on record. It would be difficult to point out any other class of marine engine of the same size as that in the *Golden City*, which could achieve 17½ turns a minute, and keep it up as a regular duty. The standard of 250 feet per minute will have to be changed, and made to suit modern pistons, as the engines themselves stubbornly refuse to be controlled by any such snail-like movement.

The fastest steamers we have afloat and the

quickest working engines in factories are beam-engines, and when we want a high piston velocity we put up a beam-engine, because we get more "turns" out of them than any other kind for paddle vessels. The *Stockton*, a passenger boat between New York and Philadelphia, makes regularly 650 and 700 feet per minute piston speed; she has a beam-engine rising 50 inches diameter and 12 feet stroke. The *Jesse Hoyt*, another fast bay steamer, 480 tons burden, has a beam-engine 48 inch cylinder and 12 feet stroke, and makes regularly on an average 576 feet per minute, burning 12 tons of coal per 24 hours in so doing. The average boiler pressure is 30 pounds, and the trips are intermittent, or short—only 22 miles in length. Several are made in a day, and the average time for this distance is 65 and 70 minnutes; it has been made in one hour with ease. The steam is worked expansively, cutting off at 8 feet, or three fourths of the stroke; formerly the influx was stopped at one fourth the stroke, but by altering the cut so as to follow 8 feet, 15 minutes better time was made, while the coal consumed was only one half a ton greater.

It should be stated that the fires are kept banked all night on the fuel mentioned, and that the net consumption for 24 hours is 12 tons.

The trouble in regard to double beat valves is overstated. Such difficulties formerly existed, but are measurably overcome. Of course if the metals composing the valves and chest are alike the ex pansion will be the same. For running in fresh water the valves and chest are generally made of cast-iron, for little or no corrosion takes place; but for marine engines the case is different, and brass must be used for the valves and seats. It was formerly customary to use disks wholly of brass for the valves, which were bolted to columns in form like spools. This practice is now obsolete in the best shops, and the "spool" is very greatly enlarged, so that it is nearly the size of the valve itself.

In an 18 or 20 inch poppet valve, for we have some of this size, the brass seat is not more than an inch thick in the average, considered through the diameter of the valve. The difference in the rates of expansion is therefore very little. It may be here remarked that most of the complaints from poppet valves arise from defective workmanship. Most engineers and latheman fancy it is an easy thing to make a pair of poppet valves, whereas there is no detail that requires nicer adjustment and closer attention. In shops where beam-engines are built, one man is kept on this work continually, and he soon acquires great proficiency in it.

Where the valves are taken apart in order to get them out of the chest, the exhaust valve for instance, it very often happens that the engineer is at fault and not the valves; for it is an easy matter to throw one of the disks out of line with the other by screwing up the bolts unequally or allowing dust or dirt to fill in.

---

## CHAPTER XXV.

HOW TO SET A SLIDE VALVE—TO FIND THE LENGTH OF THE ROD—AN IMPROPERLY SET VALVE—LEAD—THE LEAD INDICATOR.

In all the works on steam-engines which have been written we do not remember to have seen any account of the manner in which a slide valve is set, and we have had frequent inquiries from young and—must we say it—old engineers, who confessed they did not know much about it. It seems strange that any person should have charge of a steam-engine and be unacquainted with this simple duty, yet it is a fact indisputable. Many an hour locomotives have stood on the track helpless from the slipping of an eccentric which the driver was unable to replace, and mischievous comrades have oftentimes designedly loosened set screws (in the early days, when screws alone held the wheel in place,) so as to cause confusion, and subsequent dismissal, to the incompetent driver who could not reset it.

There is indeed no lack of rules in engineering works which direct us to set the eccentric, something in this way:—

"Place the crank in the position corresponding to the end of the stroke (why not say on the centre?). Draw the transverse centre line answering to the centre line of the crank shaft on the bed plate of the engine, or on the cylinder, if the engine be direct acting, describe a circle of the diameter of the crank pin on the large eye of the crank, and mark off on either side of the transverse line a distance equal to the semidiameter of the crank pin; from the point thus found stretch a line to the edge of the circle described on the large eye of the crank, and bring round till the pin touches the stretched line. When the crank is thus placed at the end of the stroke, the valve must be adjusted so as to have the amount of lead or opening on the steam side which is intended to give at the beginning of the stroke, and the eccentric must then be turned around upon the shaft until the notch in the eccentric rod comes opposite to the pin on the valve lever and falls into gear; mark the situation of the eccentric, and put on the catches in the usual way, etc."

This long and incomplete instruction is from Bourne's Catechism of the Steam-engine, and we are sorry to say omits one very important thing, so that it would be impossible to set a valve by this method. The omission is in getting the length of the eccentric rod at the outset. Without further criticism or discussion, we shall explain how an eccentric is set.

Presuming the proportions properly made by the draughtsman at the shop the first thing is—

### TO FIND THE LENGTH OF THE ROD.

Put the straps on the eccentric and connect the valve gear as in working order. Disconnect the engine and slip the eccentric around on the shaft and observe what takes place in the steam chest. Doubtless the valve will uncover one port clear to the exhaust, while the other is entirely or nearly shut. This shows the rod to be too long or too short as the case may be. If the port nearest the crank, in a horizontal engine, is wide open and the other port shut, the rod is too long and must be shortened half the difference only. We say half the difference, because it must be remembered that

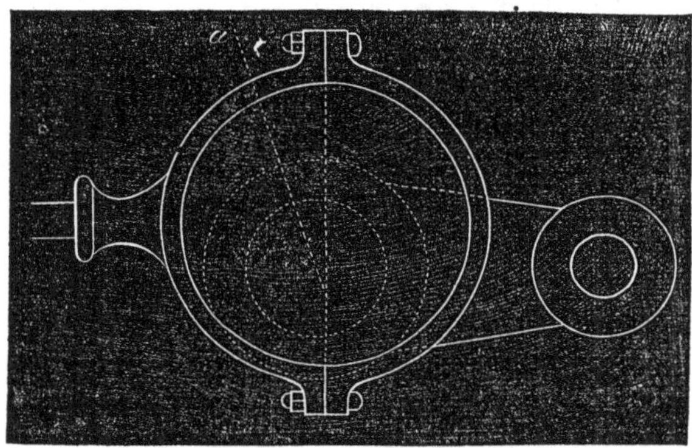

Fig. 76.

what is taken off one end is put on the other, so that the real amount the rod is shortened will be seen in a complete revolution.

When the valve "runs square," as it is called, or opens and shuts the ports properly, set the wheel as in the previous diagram.

The eccentric is always in this position in every instance, whether the engine be vertical, horizontal, or inclined. The wide part of the eccentric and the crank are always at right angles to each other, excepting such departure from a right angle as the lead and lap takes off.

The diagram represents an eccentric without lead working a valve without lap. Such a coincidence seldom obtains in practice, and the true position of the eccentric is shown by the dotted line, a; this indicates that the eccentric is turned on the shaft from the crank, thus pulling open the port in front, and driving the crank in the direction of the arrows.

It will be easily understood *why* the eccentric is always in this position, when it is borne in mind that the eccentric must commence to open the valve a little before the crank gets to the centre. In other words, the eccentric must commence its stroke a little ahead of the crank.

## AN IMPROPERLY SET VALVE.

Here is a drawing of an improperly set valve. It is not drawn to scale but is none the less a correct example. It will be seen that the crank has

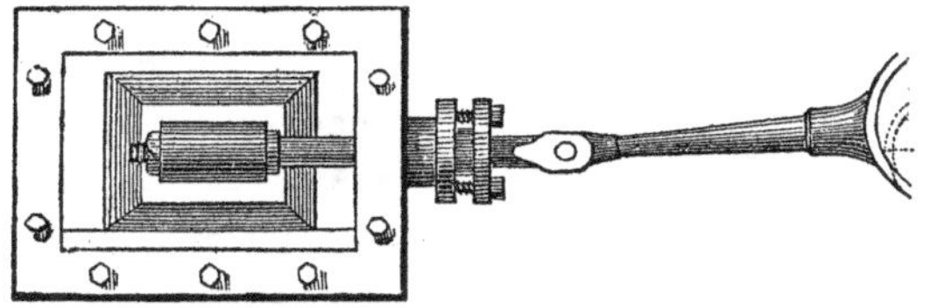

Fig. 77.

passed the centre and commenced the return stroke, but there is no lead on the steam side in front or at the port nearest to the crank, and before the crank passed the centre there must have been much compression in the cylinder at the forward end of the stroke. The steam was shut up in the cylinder and its tension or elasticity greatly increased thereby. Steam, like other gases, follows a law discovered in some experiments by a French philosopher called Mariotte. According to this authority if steam at 60 pounds be shut up in a cylinder 6 inches long, and the piston in said cylinder be pushed down to 3 inches, the volume will be reduced one half and the pressure will have been raised to double or 120 pounds.

So when the exhaust closes too soon, say at six inches from the end of the stroke, when the crank is on the centre the pressure will be in proportion to the amount of cushioning or compression.

There are many engines working in this condition to-day. Should they not be attended to?

Setting the valve with a link motion is precisely the same operation; the eccentric stands exactly as shown in the diagram. There is only this difference—the lead is somewhat disturbed by the action of the eccentric rod, which is not in gear, whether it be the forward or back connection. This derangement causes some change in the lead when cutting off at low grades of expansion; and it is necessary to take this into account when setting the valves. The lead should be given properly on the point the engine is to work at, for since the lesser rates of expansion are only used on emergency, it matters little whether they are correct or not.

Some steam chests are made with the bonnets cast on—a very foolish practice—so that the chests are merely hollow boxes, with the bottom out. It is impossible to see the valve or the lead here, and it may be set separately, and the chest put on afterward, by breaking the connections and using circumspection in putting them together again, so that nothing is deranged by false measurement.

---

## LEAD.

A point discovered in some experiments lately made in New York, is the absorption of the lead of the slide valve by expansion of the valve and cylinder, and springing of the rods when at work. Valves are set when the engine is cold, and the change which takes place subsequently is sufficiently large to affect the character of the work very much. The valve on the engine in question was set with one eighth of an inch lead, but it was found that when actually at work it opened an eighth of an inch too late, although the parts were as strong as usually made for engines of the class under trial. No reliance could be placed on the indicator to test the actual time of opening and closing of the valve, and an engineer (Mr. W T. Selden) contrived a novel indicator, which caused the valve to register the lead accurately, so as to determine the loss between the two conditions of the engine—that is, when hot and cold.

---

## THE LEAD INDICATOR.

This indicator was constructed as follows:—A wooden cylinder, covered with paper, was secured to the cylinder head close to the valve stem, in such a position that it revolved with the main engine shaft. This cylinder was equal in circumference to one stroke of the engine, so that a line traced upon it would represent in length the travel of the piston. A pencil-holder was fixed near the cylinder in such a manner that it could be thrown in or out of contact with the paper—the same, in fact, as an ordinary indicator. The pencil-holder had, further, a wedge-shaped spur on one side, and the valve stem had two such spurs, which were fixed at the lead points of the main valve. The pencil-holder was nearly in line with the spurs on the stem, so that the one on it and those on the valve stem came in contact slightly when moved past each other. It is easy to see, therefore, that when the main cylinder is rotated by a line from the engine shaft the pencil will draw a straight line, except at the lead points, where the valve stem spur and that on the pencil-holder come in contact, when a sharp, triangular break will appear in the line. The original lead line is traced when the engine is cold, and, to be a verification of it, the line should appear the same when actually running; but, as before stated, the difference was very marked.

This is a simple and beautiful instrument for the purpose, and, as it is cheaply constructed, it should be on every engine, since the time of the opening and closing of the valve are as easily seen as one's face in a mirror. The common indicator exhibits only the apparent time, while this apparatus shows the real time. Since the value of expansion, according to the law of the celebrated and immortal Mariotte, of whom so much has been said lately, depends wholly upon the extent of it, it will be seen that the lead is an important element in computing the actual volume of the steam—versus the apparent volume. It is also important as regards the mechanical action of the steam-engine, for shafts have been screwed up too tight in their bearings, pillow-blocks shifted, and connections keyed up, with a view to stop thumping, which was caused entirely by the valve opening too little, or at improper periods.

## CHAPTER XXVI.

### DEFECT IN STEAM-ENGINES.

ZEALOUS professors of science occasionally call attention to the fact that steam, as a motor, costs much more than it should, and that little over one tenth of the actual heating value of the fuel is realized in practice. Experiments and experience prove the statements to be virtually correct, and it is a reproach to the mechanical skill of the period that it should be.

The loss is not in the theory of the engine, for that is perfect, but in the practice of that theory; or, in plain terms, in the construction of steam-engines. It is an undeniable fact, however, that but few of the steam-engines now constructed work with the economy that they should, or even approximate in performance to the theoretical value of the fuel.

Portable engines are turned out by scores which, although well enough externally, are far from being in a healthy condition in those parts which affect economy. The slide valves are only such in name; they exercise few of the proper functions of this most important detail, and the boilers are heavy, enormously large in fire and heating surface, and every way disproportioned to the size of the cylinders. The feed pumps are poorly got up; the valves lift too much; the water passages are cramped and crooked, and the absence of any proper method for heating the feed water without creating more loss from back pressure on the piston than is gained by injecting hot water to the boiler is often noticeable. We make these statements for the interest of any whom it may concern —not to find fault. Many stationary engines are in precisely the same condition.

It is not the only thing required in a slide valve that it shall open and close the ports at a certain time, but that it shall be properly set for the work it has to do, that it shall exhaust the contents of the cylinder at the proper time, that it shall close properly, and that the lead shall be proportioned to the duty. That this is important every one is aware who has ever inspected, or is familiar with, indicator diagrams.

It is a common thing, on railways, to hear a

locomotive exhausting "one-sided," as it is termed, or giving palpable public evidence that it is out of order, and that the master-mechanic on the line is either indifferent or careless of his duties. We know of one road where our ears are daily saluted by the sound of a locomotive drawing a long train of coaches, and regularly exhausting 1-2-3—4, 1-2-3—4, or with a very positive interval between the successive exhausts. It would be quite as sensible to draw two or three empty coaches, day after day, as it is to permit an engine to run in this way; for at every uneven or irregular interval, the steam is compressed or choked in the cylinder, and delayed in getting out until it acquires a high tension, so that the actual pressure is much greater on the exhaust side than on the steam side. This subtracts from the efficiency of the machine, adds to the cost of repair, of fuel, and every thing used in running the engine. A locomotive engine, exhausting unequally, carries dead weight, which costs a great deal to keep.

We know that engines are often regarded as in chronic or incurable difficulties, because some mysterious cause conflicts with setting the valves properly, but we have frequently found that individuals were more fond of declaring that the defect was very mysterious, than they were zealous to remedy it.

It is very plain, from the simple facts here cited —many of which are so well known among professional engineers as to be truisms—that one of the greatest obstacles in the way of economy in the steam-engine is a want of mechanical accuracy in construction, erection, and oversight; and that the cost of a horse-power could be very much reduced by attention to obvious and well-known defects existing in steam-engines.

## CHAPTER XXVII.

### THE SLIDE VALVE—BALANCED SLIDE VALVES.

THIS most essential detail of a steam-engine is very often badly constructed, set, and run. The valve may be called the heart of the machine, and any derangement in its functions results in loss of money, power, and reputation of the builders, and all concerned in running or erecting steam machinery. In many places we have noticed a disregard of the commonest principles connected with the designing of slide valves, and deem it our duty to point out some frequent errors, so that they may be detected and rectified.

When the lead on the steam side of the valve is open, the exhaust side is closed, and the steam behind the piston cannot escape until the valve has travelled far enough to open the exhaust port, which is a greater or less distance according to circumstances. This is one and a very serious defect; a piston is not meant for a punch, and steam is of so subtle a nature that, give it but the slightest opening and it will rush through like lightning. To remedy the evil just mentioned, take the steam chest entirely off the cylinder, if possible, take up the valve, and with a square and a scriber mark off the width of the faces which cover the ports on the outside of the valve; pursue the same course with the ports on the cylinder; then replace the valve, make the connection with the valve-stem, and turn the crank on the centre; the relative situations of the steam and exhaust ports will then be apparent at a glance if the eccentric is properly set. The distance or amount of opening which is proper on the exhaust side of a slide valve varies with the effect desired to be produced, and also with the ideas of different engineers; some claiming that a small amount of lead should be given to the exhaust, so that a portion of steam will be retained in the cylinder for the piston to cushion against: thus producing an elastic vapor, which reacts to advantage when the cranks are passing the centre.

In locomotive or high speed screw-engines, for instance, a certain amount of compression at the end of the stroke is essential to the safety of the machinery. The theory of compression is, however, a dangerous one, especially to novices in engineering, who are liable to overstep the bounds and cause loss where they intended gain. There is much more benefit to be derived from a clear field for the piston, or from the partial vacuum which is obtained through large exhaust passages and properly set valves, than in all the fine-spun theories about cushioning, filling the passages with steam, etc.

In designing the outward form of slide valves there is a great deal of carelessness exhibited respecting the amount of surface exposed to the action of steam. Fillets are made unusually large, flanges extended unnecessarily, and extraordinary lap introduced, until the aggregate value of all the useless surface amounts to an addition of many hundred pounds pressure on the valve, when the steam worked is of a high pressure. The line of contact of the seat and valve, or the two faces of the same, should be as accurate as possible, and this detail requires close attention in order to make the valve work with economy. After an engine has been running for some time the seat acquires a glazed surface, which is very difficult, if not impossible, to cut with a file or scraper, and the proper way is to make the valve and seat true at first, and not trust to its wearing fair in time, although this method is often practiced. The valve should be surfaced true by the aid of a metallic face plate, where it is possible, and the seat should then be scraped from the valve. When the valve is put into the chest, the faces of both it and the seat should be carefully cleaned with a pocket handkerchief, so that no grit or dust, even, can possibly remain upon either; as the smallest particle will in a short time ruin the faces by working seams or ruts through which the steam leaks. The balancing of slide valves should also be attended to; a portion, at least, of the pressure might be taken off with advantage, and the mechanical effect would be much increased thereby.

---

## BALANCED SLIDE VALVES.

A serious objection to the use of slide valves is the great power absorbed in operating them. When this detail of the steam-engine is of any considerable size the evil increases to an injurious extent, so that the force required to work the valve is a large percentage of the whole power of the engine. Indeed, it is not unusual to see slide valves which have an area closely approximate to the total number of square inches on the piston, and if the nature of the work to be done were similar in both cases, the pressure on the piston would be balanced by that on the valve, and the engine would not move.

The result of this great labor on one part of the engine is followed by the rapid wear of every thing connected to or with it. The eccentric straps of screw propellers are often made of wrought-iron and bushed heavily with brass, so as to withstand the strain and to avoid heating caused by the excessive friction. The rock shafts are made unusually heavy, the valve stems much stouter, the journals larger and longer, and in fact each detail is very greatly enlarged so that it may be capable of performing the task assigned to it. Thus a larger amount of material is used than is needful, an increase in the cost of construction is apparent, expense in lubricating and repairs ensues, and the whole system is not only defective from an engineering point of view, but vexatious in its commercial aspect.

Aside from this defect, the slide valve, when properly made, is one of the simplest and most effective devices ever invented for its office. If there were no remedies for the disease spoken of previously, the adoption of the slide valve, beyond certain areas, would be discouraged, but since the ingenuity of man has provided a way of escape, it is singular that so few interested parties avail themselves of it.

If we were temporarily made a power from which there could be no appeal, we should immediately fulminate an edict against parties who use slide valves, and command them, on pain of large annual repairs, and manifest deterioration of their property, to apply some method whereby the pressure of the slide valve would be reduced to a rational amount; no greater than that due to the work required of it.

When the first invention to relieve the excessive friction of the valves was brought out it was a step in the right direction; but the attempts to introduce this valuable improvement have met with very little encouragement from those most interested. In the cases of large screw propellers which have engines working at high speeds it is absolutely necessary to divide, or take off the pressure on the valve face. Different methods have been adopted to do this. One of the simplest for low-pressure engines is that invented by Messrs. Penn, of England. This consists in planing the back and face of the valve parallel and placing a brass ring between the back and steam chest bonnet, so that the junction is steam tight. This ring covers only a portion of the area of the back, and therefore excludes the steam pressure from that area. A connection is maintained with the space inside of the ring and the condenser, which materially aids in restoring the balance, or reducing the friction of the valve on its face.

A striking effect of the utility of this contrivance, so far as relieving the pressure is concerned, was witnessed by engineers on the Italian frigate *Re d'Italia*. The condenser communication was shut off, when the valves, stems, and rods, although of ample dimensions, trembled violently, showing that the resistance to motion was very great. On restoring the connection the valves resumed their previous easy movement.

Another method is to attach a steam cylinder to the steam chest. In this cylinder there is a piston which the main valve is connected to by links; when the steam is let into the chest it presses equally on the valve and the piston, between the two, so that the pressure is taken off the valve in the ratio of the difference between its superficies and that of the piston. The piston has a slight reciprocating movement to compensate for the stroke of the valve. Vacuum may be maintained on the back of the piston so that by the combined force of the steam and the absence of back pressure on the piston the valve may be actually pulled off its seat. It is impossible to detail all the ingenious and practical devices for this purpose, but it is manifestly proper that they should be universally adopted.

## CHAPTER XXVIII.

### CONNECTION OF SLIDE VALVES—THE PRESSURE ON A SLIDE VALVE.

THE essential virtue in the mechanical adjustment of a slide valve is that it shall open and close the ports at the proper time, and that it shall be steam tight. Other considerations present themselves, such as the proportions, friction, etc., but we confine our discussion to the connection between the stem and the valve itself. A slide valve may be properly fitted to its bearing; but by reason of a badly designed or applied connection with the stem, be rendered inefficient. How many of our readers experienced in these matters are there who have not noticed that the slide valve is (oftener than otherwise) worn winding, or all on one side, when there was no apparent reason for it. The cause can generally be attributed to the stem and its connection. Let us examine the ordinary plans in use for working a valve. If we do so, we shall find that the form generally employed is a simple nut, in which the stem is screwed, fitted into a pocket on the valve. This kind of connection is in use on some very large engines, and it is not at all to be commended. The stem working through the stuffing-box, has a very material vibration, and does not by any means work in a straight line. The packing affords no protection whatever against the evil, and the stem may deviate measurably from travel in a true line, to the manifest injury of the engine.

The supposition is that the nut, being easily fitted, will give a little, up and down, and let the valve work fairly on its face. Such is not the case, however; and the stiffer the valve stem is, the greater the evil. It constitutes a lever which works on the stuffing-box as a fulcrum, and pries the valve up so much that it wears harder in one place than another. The pressure of the steam is not sufficient to overcome the strain exerted on the valve stem by the several connections. Even when guides are provided, the same evil is not wholly obviated: as they are not always set in a direct line with the valve face.

Another popular form of connecting a valve to its stem is found in the square yoke, fitting completely about the upper part of the valve, and in some cases providing it with a tail which runs through the back end of the chest. The double stuffing box is a good feature, as it insures a true linear movement of the valve stem; or at least one more correct than is ordinarily obtained. But without this provision, the valve is even more liable to tilt than with the single nut; for the reason that the surfaces in contact are greater. Slide valves are also driven by a nut laying in the centre of the top through which the stem passes. This is perhaps the best form of applying the stem for general use; as it insures a direct pull from the centre of the object moved, and does not create an undue twisting or straining of the valve itself. Too often the face and seat of a valve seem to indicate a true surface by their polished appearance; but upon examination by proper instruments it will be found that they are not so. The slide valve, as a means of controlling the energies of the rest of the machinery, should be carefully and frequently examined to see if it is in perfect order, as much loss results by its imperfect action. A leaky valve destroys not only its own face, but that of the cylinder also; and the latter is renewed only at an expenditure of much time and labor.

---

## THE PRESSURE ON A SLIDE VALVE.

It is a popular idea that the number of square inches in the back of a slide valve, and the pounds of steam in the chest, represent the total pressure upon the valve. Another delusion is, that the pressure on a slide valve is equal to the pounds of steam per square inch on the back, minus the area of the steam ports. If we consider the valve to be a solid block of iron on a solid table, and mechanically tight, the steam would press on every square inch of surface with the same force that a dead weight laid upon it would. But these conditions are never found in a slide valve, except in one position; that one, when the valve laps over both ports, and the engine is at rest.

So soon, however, as the valve is moved the steam enters the open port and the pressure is practically taken off that end of it. When the valve is moved back over the port, the steam that is shut up within the cylinder will press up against the under side of the valve face with a force exactly equal to the pressure at the point in the stroke of the piston at which the valve closed. As the valve continues its stroke the other port will be opened, and the steam we have supposed shut up in the cylinder begins to exhaust; at this time, the pressure against the under side of the valve will be the pressure in the cylinder at the end of the stroke. This pressure is only for a brief period, however, for in a well constructed engine the time of exhausting the contents of the cylinder is very short. While the steam is entering the open port, then, and after the exhaust has passed through the closed port, the pressure on the under side of the valve will be just the ordinary back pressure, supposing the engine to be non-condensing—which is the supposition we have entertained in this discussion.

It is therefore unquestionable that to determine pressure on a slide valve we must consider the pressure in the cylinder at the time of cutting off, at the end of the stroke, the area of the ports, the area of the back, and the back pressure on the piston.

## CHAPTER XXIX.

### CONDENSATION OF STEAM IN LONG PIPES.

SOME information, exceedingly interesting to engineers, has recently been made public in an account of a subterranean engine erected in the celebrated "Gould and Curry" mine, California. The engine is 50 horse-power, and is 201 feet feet below the surface of the ground. The experiments were made in 1865 by Mr. R. G. Carlyle, Chief Engineer.

He says: "The mine was worked through three tunnels—upper, middle and lower—with a respective difference in their levels of about 225 feet each. In consequence of a very heavy winter and the softening of the hanging wall of the mine, it became evident that the mine would cave or fall in; therefore it became necessary to project some other works which would secure the yield of the mine at a lower depth, outside or below the 'cave.' There was no shaft from the surface, so that there had to be put up temporary works in some secure part of the mine until a shaft could be put down from the surface. I then carefully considered the troubles arising from putting a boiler in the mine; and, on the other hand, the ease with which a steam pipe could be carried there from a boiler on the surface. In fact I had no other recourse, as, if I put a boiler in the mine, I would have to use part of the old workings for a smoke stack, but as that was going to 'cave,' I would then have had no smoke stack at all, so I resolved to carry the steam 1,300 feet, which was the shortest available distance to the surface. I had no data to work on other than the knowledge that, in some coal mines in the north of England, they have carried steam six or seven hundred feet for accessory work, from lower levels than the main pumping level. It was 'Hobson's choice' with me, but I was fully aware that I staked my reputation in the experiment.

"The boiler was of the common Mississippi style—two flues of 42 inches diameter 26 feet long, and two flues 14 inches diameter, having also steam and mud drums. The steam was taken from the steam drum and passed through a superheater under the boiler—the same firing answering for both—and thence through a 4 inch gas pipe down an air shaft to the lower tunnel, where I had fixed an expansion joint and also an accumulator; this was a small boiler, 30 inches diameter and 5 feet long—its object being to catch water in case the boiler should foam, or to drain the pipe beyond. As the pipe raised gradually from this accumulator to the engine, with the grade of the tunnel, it was in just the right place. The length of the steam pipe in the air shaft was 201 feet. From the accumulator the pipe ran alongside of the tunnel, to a branch tunnel, to the engine room—600 feet long--in the branch tunnel—500 feet long—and up a slight incline to engine room, 40 feet more—making, in all, a steam pipe of 1,341 feet in length. In the engine room was placed another accumulator, the same as the one at the bottom of the air shaft, but set on its end—the steam going in at its middle and out to the engine at the top. The object of this one was to catch whatever water might be carried with the steam, also scale from the iron pipes, and to form a kind of reservoir for steam; as the engine had a variable cut-off on, it acted as such to a considerable extent. On each of the accumulators, was placed one of Furman's steam and water traps, also a gage to note pressure.

"The engine was made at the Vulcan Iron-works in San Francisco, and was a horizontal cylinder of 14 inches bore, 30 inches stroke, and was used cutting off a half stroke. It hoisted a bucket for sinking purposes, holding one ton of rock, in one shaft 200 feet deep; in another shaft a cage, with car and load weighing 3,000 pounds. The speed of hoist was 400 feet per minute; it also worked a pump of 8 inch bore, 4 feet stroke, with its machinery in the third shaft. The amount of water was not much—about half the capacity of pump, as the pump was going sucking about half the time. The trips of hoisting were made about every ten minutes, respectively—sometimes both were hoisting together. The hoisting apparatus was of the friction variety—the same as generally used in these mines; in all I think the engine had to do about 35 horse-power of work.

"The steam pipe was 4 inch gas pipe screwed together with flanges at intervals of 100 feet. For convenience of repairs, in every 400 feet there was an expansion joint. The pipe was anchored to the side of the tunnel in the middle of that distance, so that it expanded both ways from that point. The casing of the pipe was of wood, made of two by 12 inch plank—making a box of eight inches square inside, in the centre of which rested the pipe on saddle pieces, the balance of space being filled with common wood ashes. The expansion of the pipe was very nearly two inches per 100 feet, from 60° to temperature of the steam at 80 pounds pressure. [325°.] The difference in pressure at the boiler from that at the engine, could not be detected; I changed the gages (Ashcroft's) from the boiler to the engine, but no difference could be found. I even made two gages of gas pipe, half-inch, of common siphon shape, and filled them with mercury. I made them long enough to suit our working pressure, and still no difference in pressure between boiler and engine. I also made experiments without the superheater, and found no difference in pressures. The only loss was an increase in the amount of water trapped off from the pipes. The loss would then be one cubic foot per hour trapped off; with the superheater the loss was one third of a cubic foot per hour. The amounts trapped off were accurately kept; these figures are the average, and not the result of any one hour, although it never varied much from what is given. When the flow of steam through the pipes was rapid it was less; when slow, greater.

"The fuel was common pine wood, using from three and a half to four cords per twenty-four hours—which will compare with any engine having short steam pipe and doing the same amount of work with the same kind of fuel. The engine ran in the mine over one year, during which time I made numerous experiments with it. It is now out of the mine, as they have no use for it in there. It was a complete success, as it did more than was ever expected of it, and enabled the company to declare dividends during the 'caved' condition of their mine.

"In conclusion, I would state that, as far as my experiments went, I see no end to the distance to which steam can be carried—it being merely regulated, more by the amount of condensation than by difference of pressure. I would not hesitate to carry it one mile, if I could cover the pipe well—that being the great point to be looked after."

## CHAPTER XXX.

### PACKING STEAM PISTONS.

ECONOMY in fuel, saving in repairs, in short increased duty generally, results from well-packed steam pistons. Erroneous views respecting the performance of this duty prevail among engineers. It is thought necessary to use great force to insert the springs; the springs themselves have set screws in them, they are tremendously thick and heavy, entirely disproportionate to the work, and, in most cases, not what they should be. The surface of a steam cylinder is one of the most critical or nice points of the engine; when it is once ruined heavy expenses are incurred in renewing it; and since it can only be injured by gross carelessness, it behooves every engineer at all anxious for his reputation to be sure that he does not omit any portion of his duty toward it.

When the piston has been removed from the cylinder from any cause, the utmost caution should be observed in replacing it. Not only should the rings themselves be chalked before the follower is removed, so that they may occupy the same place in the cylinder, but every minute speck of dust, and the little concreted balls of tallow and sediment, which collect like shot in the inside of the piston, should be taken out and the scale scraped off. Sometimes the heat and tallow combined cause a thick, heavy deposit to appear on all parts inside the follower; all this matter should be removed and the piston rendered as clean as it came from the shop, if possible. When the rings are inserted they should be wiped with something clean and soft, so that there can be no possibility of dirt or grit adhering to them. The piston should be put in its place, and if it belongs to a horizontal engine, the rod packed; in a vertical cylinder this is not possible. When the piston is carefully centred by means of inside callipers, the rings should be pushed in by hand, and the springs, properly "set," driven in with a tap of the hammer handle. In horizontal engines the weight of the piston must be compensated for by extra strong springs or blocks of wood; but if the cylinder is true and has been well taken care of, the springs correct in proportion, and the rings of a proper thickness, the piston will be perfectly steam-tight and easy working with the packing set out as described. An immense pressure is exerted on the interior of the cylinder by stout springs, and friction is generated to an alarming degree.

Let any engineer take one of the springs usually employed in horizontal engines, and place weights on it, and see how much it requires to deflect the centre one sixteenth of an inch; he will then have some new ideas about the friction of the packing in an improperly-packed cylinder. The barbarism and absurdity of using set screws in packing is too manifest to call for criticism; such devices are not springs, they are small jack screws, and have no more elasticity than a solid block of iron. We once saw an engineer take an oak stave, as heavy as a stick of cord-wood, and batter on the end of the piston so as to drive it in, the packing and springs (which he had previously inserted on the floor to save time), into the cylinder. It is not necessary to remind sensible men that such a course as this is simply that of a blockhead. The packing of a steam piston should be examined once a month at least, to see that the springs have not set or relaxed, and that every thing works well. If the rings are too slack edgeways, the follower must be "skinned" over in a lathe, so as to bring the surface of it and the packing in contact again. Much care should be taken that the follower does not bind the rings tightly. When the bolts in it are screwed up hard, the rings should be so fitted that they will slide in and out with a strong pressure of the hand; then the packing will perform its functions properly.

# CHAPTER XXXI.

## PISTONS WITHOUT PACKING.

WHEN the first pistons to steam-engines were made they were made tight by hemp gaskets—that is, coils of hemp plaited with rope thoroughly slushed or soaked in hot tallow, and subsequently driven in as tight as a man striking with a sledge could make them. It was a great step in advance when cast-iron rings were substituted for the hemp and steel springs inserted to keep the rings always up to the cylinder. Quite as much ingenuity and thought have been expended on the pistons of steam-engines as upon any other detail, and the variety in shape, form, and kind of packing would make an interesting study for the engineer if they were all collected in book form. The pistons of ocean steamers, for instance, have lighter springs than many small engines, and are not packed so tight, by many degrees pressure, in proportion to their areas, as some engines on land. There are few stationary engines in the country which will pass the centres with two or three pounds pressure on the gage, but there are plenty of steamboats that have engines which will do this with ease.

It was formerly the custom to pack locomotive

cylinders with brass rings, which had a central lining of Babbitt metal let in. This also is done away with, and the largest works and the heaviest engines on the Erie Railroad, and others, for aught we know, have cast-iron rings.

In many instances pistons have been used without any packing in them—being simply solid disks fitting tightly, yet easily, to the bore. Some concession has been made to prejudices and conventional ideas by turning grooves in the solid piston and depending on the partial condensation of the steam to fill these grooves with water, and thus interpose an obstacle to the passage of steam between the piston and cylinder. It is probable that the evil of a leaky piston has been much exaggerated, for, although it will show on the indicator diagram when very much out of repair, it is a question whether any great amount of fuel is wasted by such a loss. There is no question, however, but that much damage is done to steam cylinders by bad packing, and many can testify to the scored and seamed cylinders that were made so by forcing in the springs.

Air-pumps have been made for compressing air with solid pistons, and, reasoning from analogy, there seems no objection to making the pistons of steam-engines of a moderate diameter of cylinder entirely solid; in fact, many are now working so made, and those who built them, as well as the owners, find no fault with their performance. On the contrary, rings are frequently a source of trouble, and, taken altogether, with their springs, followers, and follower bolts, the piston with metallic packing is a costly detail. If lessening the cost of construction, and retaining the vital qualities of any part, is an important feature, then the pistons of small steam-engines should be made solid.

## CHAPTER XXXII.

### BEARING SURFACES.

THE economical working of machinery depends upon many things—the care observed in using it, the material employed in its construction, and, lastly and chiefly, the proportions of the design; for good workmanship, material, and careful supervision may, for the purposes of discussion, be assumed. The resistance of every machine is very greatly increased or diminished according to the harmony of proportions existing between the several principal parts. The wear of the shaft, the burthen on the beam, the wear and tear of cylinders and packing rings, the weight borne by the guides in sustaining or directing the cross-head, all these points have some importance in the general economy of a steam-engine. So also does want of proportion affect the performance of other machines when transferred to them, and the best test of durability, and, as a sequence, economy, is found in engines which have run for years without repairs—equal engineering skill and similar conditions being assumed for the purpose of comparison.

If we examine the V-shaped slides of a planing machine we shall find that they do not wear equally, considered through their cross-sections, and that in most cases the points which wear the most are nearest the top of the slide or at its apex. The base on such angle is always the brightest, showing that the most friction occurs at those points. One reason for this may be found in the shape of the wearing surfaces; the form is so unfavorable to lubrication that oil will not remain upon it very long, but runs down toward the lowest parts, carrying with it the dust that may have settled on the slides. Instead of making the slides in this way, it would seem a better plan to take off the top of the triangle (considering the slide through its cross-section) so as to transfer a portion of the wear which the lower parts of the inclined sides sustain, to a flat or plane surface. By this method of construction, which is often practiced on the shears and lathes, the wear would be more equal and even, and the slides would last longer without replaning. Many makers of planing machines extend the base of the slides very much, so as to make wide and heavy bearings, and this plan has been found to answer well for large machines. We once saw a planer with slides which were semicircular or rounded on the top, and they worked very badly indeed. The sides of the semicircle, if we may use such an expression, wore off much quicker than the top, and the consequence was that the surfaces in contact never fitted.

The journals of steam-engines are very often made convex in their axial length, some are made concave, and still others have coned bearings to certain parts. These plans are all defective for these reasons:—the wear is unequal because the velocity of the surfaces in contact is unequal; the pressure upon the bearing is not the same through out the surface; the lubrication is imperfect, because the oil flows from the highest to the lowest points, so that in a short time the greatest diameters are left dry unless more oil is poured on than a journal of similar size properly made should have. Any departure from a true cylindrical surface is costly to manufacture, while the use of such journals is not attended with advantages sufficiently great to counterbalance their evils except on traction engines, some parts of quick-working screw-engines, or places where great strain is liable to be thrown on the parts connected—as in long connecting-rods or self-propelling engines for common roads.

Quick-working screw engines, having short

strokes, with the crank shaft so near the cylinder-head that it makes the latter squint-eyed to look at it, wear down their gibs in the cross-head (when they have any) most rapidly, and no remedy for this appears to exist but to make the gibs either of hard wood boiled in linseed oil or else brass, disproportionately large for the area of the piston. Wooden gibs wear while the slides do not, which is a very important advantage. We have seen the gibs of a cross-head belonging to a direct-acting vertical screw-engine (said gibs being of brass, about 14 inches long by 8 inches wide, worn down nearly three quarters of an inch on their face in going from this port to Savannah, Georgia, in spite of all the oil that could be poured on, or attention that could be given them; it may be proper to state that the cylinder was about 50 inches in diameter and the stroke 48 inches. As an economical substitute for small brass boxes, lignum-vitæ boiled in oil or tallow is very good, and is used to some extent in many quick running machines.

# CHAPTER XXXIII.

## LUBRICATING THE STEAM-ENGINE.

So much waste is continually going on in the very costly item of oil that we are prone to think the cause of it is ignorance and not wholly carelessness, for the most reckless person would hardly be guilty of such criminal waste as is too often manifested. In the matter of supplying the furnaces of a boiler with coal there seems to be little hope of radical reform, for the quantity of half-burned fuel, clinker, and needless waste we have seen thrown out with ashes is astonishing.

Oil is not a motive power, and possesses no impulsive energy whatever; but to judge from the manner in which it is slopped about, one would suppose that it had some peculiar virtue hitherto undiscovered. Moreover, the amount of oil which a bearing will carry is limited, and after the surface of it and the brass is once covered, every drop poured on is wasted, for the journal throws it off, just as the stomach does food when overloaded. The proper quantity for any given bearing can be learned only by experience, and engineers will find that they can economize very greatly by a little observation and practice. In addition to this test there are automatic or self-regulating devices for limiting the supply of oil to bearings. These consist of oil cups, but the use of them if not abused is greatly misunderstood. An oil cup is not simply a funnel to pour oil into so that it will run down on a bearing, but it is intended for a feeder, or to gage the quantity supplied as exactly as a cock admits water to a boiler. This function is performed by one of the most simple arrangements it is possible to conceive of—namely, a length of cotton wick and a tube. The tube is always fitted in every properly made oil cup, and the wick only remains to be supplied by the engineer. No engine should be without a cup on every bearing, for it is impossible to judge as accurately what the journal requires as the wick will feed when once properly adjusted. The tube in the oil cup must not be filled too tight with the wick or it will not feed, and by regulating the size and fit of the wick the oil may be fed fast or slow. Besides the wick there are many other self-feeding oil cups working on scientific principles, which perform very well, and we have merely alluded to the simplest one—that with a wick, as an example of a self-feeding arrangement.

It is not alone the exterior working parts that require lubrication; but the valves and piston should occasionally receive attention as well as the others. We are well aware what a disputed point this is in engineering practice, and can ourselves bear witness to very many steam-engines that have been running for years without a drop of oil in their cylinders or valve chests, but it must be borne in mind that these are exceptional cases, and may be due to the state of the steam, the construction of the engine, and nature of the metals in contact. By the state of the steam is meant whether it is high-dried or superheated from the construction of the boiler, or whether it is moist, as occurs in boilers with small steam room. Therefore, because one engine here and there runs without oil in the parts mentioned, it does not follow, as a rule, that no valve-seat or piston requires to be lubricated. The medicine which is harmless if taken by one man becomes deadly poison to another, and where in one instance oil would not only be wasted but would injure the machine if applied, in the other it is absolutely essential to economy.

The frequency with which oil is to be introduced into the cylinder is a very important point, for the same rule applies here as to the bearing. All that is not essential to the work is thrown away by the engine or carried out with the exhaust. In proof of this the tubes of some surface condensers were recently discovered to be half choked up with tallow; if this fact were not substantiated by the assertion of a competent engineer we should be inclined to doubt it.

Without further digression, it is important to

observe that it is not only the waste of grease which occurs by its lavish use, but the injury which those vital parts—the cylinder, valve-face, and packing—sustain, that render the employment of tallow a source of damage to the parts mentioned. It is well known to experienced engineers that there is a peculiar appearance of packing rings, and the piston faces where they set, when much tallow has been used. This peculiar appearance may be called "worm-eaten," since it resembles nothing more than it does the track of an insect; it is to the endurance of the metal just what the ravages of the worm are to timber. In a short time it is wholly destroyed. Some pistons that we have seen might have been cut with a knife. This damage is wholly owing to the free use of grease, and is explained by these facts.

In the rendering of rough fats, such as are used for greasing cylinders and pistons, sulphuric acid is freely used. The quantity of the acid used amounts to, at the least, 12 per centum of the weight of fat, and it combines at once with the whole of the fatty matter; a portion of the acid is removed by washing the grease in water at a high temperature; but a certain part remains behind and becomes a constituent of the rendered mass. This acid is set free when introduced to the steam cylinder by the heat therein, and though necessarily small in quantity to the proportion of grease introduced at once. exerts its evil influence, and slowly but surely destroys the metal. The iron is eaten up, and the carbon alone remains. Where this adulterated or impure grease has been used too lavishly the iron is entirely destroyed, and what remains may be literally cut with a knife like charcoal. Animal fats are themselves acids, chemically speaking; but these are not specially injurious to iron. The best way to avoid this trouble is to use pure grease or beef tallow, rendered by heat alone. A quantity of this placed in a tin pot on the steam chest will gradually resolve into a pure fat, and by pressing the scraps but little or no waste takes place.

The practice of going about with a squirt can and spirting oil *at* the bearings—not into them—is a most reprehensible one; and no conscientious man in charge of machinery would do it. Oil at $1 50 a gallon is too dear to be used in this way, to say nothing of the moral nature of the act. Cups that feed themselves require to be watched to see that they do not stop feeding; and, in fact, every department of the engineer's duty requires vigilance, so that the wonderful machine may be kept up to its duty.

A capable engineer is always in request; and if any go about idle, let them ask themselves if it is not because they have not studied their interest by using diligence in the discharge of their duty.

# CHAPTER XXXIV.

## DERANGEMENT OF STEAM-ENGINES.

In a large steamship, when the engines are being placed on board, there are so many irresponsible persons about, so many different trades at work (on the same detail, perhaps), that it is a matter for wonder that accidents do not occur to the engines oftener. We have many times noticed oakum plugs, shreds of canvas, etc., stuffed into valves and cocks, to keep dirt and chips out; and we have also seen these preventives drawn in, or purposely thrust down, by mischievous individuals, so that when the cock was opened they would effectually choke the pipe. To our thinking, it is better to avoid the chance of disaster in this respect, by making a plug, say 6 inches long, and driving it in firmly; then there will be no possibility of damage. The boilers of a new steamer are usually the receptacle of every conceivable sort of rubbish. No one who has not actually seen the contents of one can form any idea of the matter that is collected in them. Old shoes, that the boiler-makers carry in to throw at each other, keg-staves by the wheelbarrow load, the refuse of the rivet kegs the men sit on while at work, hoops from the same, old lamps by the quantity, red lead, washers, nuts, bolts, rivets, chisels, hammers, spun yarn, and filth unnameable, form some portion of the miscellaneous contents of a new steam-boiler in a large ocean vessel. It is no wonder, then, that they foam when first started; all this refuse has to be cleaned out as best as it may, and much of it is never removed. When cold water pressure is applied to steam-boilers, they being filled full to the safety-valve, it is very possible that some of these sticks or staves may be carried into the stop-valves, and remain there, to be at some time drawn into the steam pipe, and from thence to the engine. There is certainly danger that before the worthless litter gets so water-logged as to lose its specific gravity, it may do some damage. A good plan to avoid danger from the cause mentioned, would be to let cold water into the boilers a few inches over the flues, and before they are fired up have some trustworthy man to go in and clear out all the foreign matter he may find. He might not discover any thing, and possibly sticks by the armful might be found: at any rate no harm could ensue, but much benefit might result from the observance of this plan. We have seen bolts and chisels taken out of steam boilers that were covered with scale, having been in them ever since the boilers were made, and know also of one instance, at least, where a new engine, when it was to be started, utterly refuse to move an inch, and the bonnet being removed from the steam-chest, a chipping hammer was found jammed between the valve and chest, probably carelessly left in after the valve was set.

Every new engine should be thoroughly inspected before it is tried in actual work, to see that nothing is out of the way; the accident proves that it is safer to err on the side of prudence than the reverse. The engine of a large Frigate became disabled, from the fact that the "pocket" in which the valve stem nut is fitted, pulled out, and the broken pieces falling down into the port were blown into the cylinder by the steam entering subsequently. The piston was much injured.

## CHAPTER XXXV.

### COLD WEATHER AND STEAM-ENGINES—OIL ENTERING A STEAM CYLINDER AGAINST PRESSURE.

DURING the winter much more care is necessary to preserve steam-engines from injury than in milder seasons. Feed pumps are particularly liable to be damaged by frost, and much delay and expense results from inattention to them. Every pump should be provided with a small cock, so that the water could be drawn off every night, and the same should be left open so that no driblets or leaks from the suction or supply pipe could run in and cause damage, as pumps are so situated that this might occur sometimes. A steam cylinder needs a warm coat in winter as much as a man does, and if at no other time of the year, the pipes and all other parts containing steam should be "lagged" or felted heavily, as the loss by radiation is something to be considered. It is no argument to say that the engine room is itself warm enough, for this is not so; heat is radiated from all bodies, whether their temperature be the same or nearly the same as surrounding bodies; for it is the tendency of heat to place itself in equilibrium. The strain on a feed pump, induced by freezing the contents, amounts to one eleventh of their bulk, as water expands in that ratio by freezing. An unloaded shell, it is said, was once filled with water and exposed during a cold day. The hole was stopped with a plug, which was thrown violently out of the shell, when the water froze, to a distance of 400 feet, while a cylinder of ice eight inches long protruded from the aperture. In excessively cold weather, where steam-boilers are allowed to get entirely cold over night and are fired up again in the morning, they will soon become leaky; as the constant extremes of expansion and contraction tend to produce that effect. An immense amount of fuel is wasted every year, even with the most careful supervision; but the quantity becomes enormous when little or no care is taken. In the winter this is particularly the case, and some steam pipes are as cold as if they had never had a pound of pressure in them; the result is easily seen at the end of the year.

## OIL ENTERING A STEAM CYLINDER AGAINST PRESSURE.

In the Detroit Locomotive Works there was at one time a vertical high-pressure steam-engine (since altered to low-pressure) which had an oil cup on the cylinder head. By opening this cup, during either the up stroke or down stroke of the piston, oil would flow in, although the steam gage indicated some 18 or 20 pounds pressure. This was somewhat remarkable. Oil would naturally

Fig. 78.

flow in on the up stroke of the piston, because the exhaust would then be open, and the pressure less than that of the atmosphere; but how was it that steam did not blow out on the down stroke instead of the oil running into the cylinder?

By watching the operation closely we discovered that the oil was drawn in during the first portion only of the stroke, that when the stroke was nearly completed the action was reversed and the oil was blown outward. Seeking for an explanation of this singular circumstance, we observed that the pipe from the oil cup entered the cylinder through the head, and directly over the steam port. We suppose that the oil was drawn in by the friction of the steam in its passage through the port.

A simple case of this kind of action is illustrated in the annexed cut. The two pipes communicate with each other, and the lower end of the vertical one is placed in a tub of water. Now, if a current of water, steam, or other fluid is forced rapidly through the horizontal pipe, it will carry along by friction the upper particles of any fluid filling the vertical pipe. The pressure of the air will force up other portions of the fluid in the vertical pipe to take the place of those removed, and these will in their turn be carried along. Thus an upward current will be created in the vertical tube. Steam pumps have been constructed on this principal. We suppose that this was the action in the case of the oil cup at Detroit.

## CHAPTER XXXVI.

EXPLOSIONS OF STEAM-BOILERS—BOILER EXPLOSIONS—IS YOUR BOILER SAFE?—FAULTY CONSTRUCTION OF STEAM-BOILERS—TROUBLES INCIDENT TO STEAM-BOILERS—STARTING FIRES UNDER BOILERS—STEAM-BOILERS AND ELECTRICITY—FIELD FOR IMPROVEMENT IN STEAM-BOILERS.

IN Manchester, England, there exists an association of engineers who carefully survey every disaster of this kind upon its occurrence, and report the prominent features which, in their opinion, were the cause of the accident. They not only do this, but they also inspect the boilers of all persons who are members of the society, from time to time, as they deem necessary, so that every reasonable chance of explosion may be anticipated, and the proper means taken to prevent it. The results of this organization are forcibly shown by the report. Out of thirty-six boilers which exploded in 1863, but one of them was under the charge of the association, and this was an exceptional case: all the others ran their chance, as we may say, and suffered accordingly.

By a tabular statement we find that the principal

cause of explosion with most boilers was corrosion—chiefly external. The report also mentions that damp, or "sweat," as it is sometimes termed, formed between the walls in which the boilers were set, and thus caused the injury spoken of.

Careful and deliberate synopses of the several disasters enabled the members of the inspecting committee to arrive at the conclusion that one sixth of the explosions which occurred could be traced *directly* to this external corrosion. From this it appears, that however important it may be to examine the interior of the boiler, it is also of vital importance to investigate the outside, especially those parts which are either in immediate contact with the setting walls, or else so covered by them as to prevent thorough ventilation.

A very general opinion prevails that explosions arise either from shortness of water, tampering with the safety-valve, or excessive pressure. An examination of the table alluded to does not warrant this assumption, for out of thirty-six explosions only two were from excessive pressure, four from scarcity of water, and but one of the cases of over-pressure was caused by carelessness, the other being an inadvertency.

---

## BOILER EXPLOSIONS.

Boiler explosions are becoming remarkably

prevalent. Scarcely a day passes but what, from some part of the country, remote or near, we receive intelligence of a great disaster. It is perhaps inevitable that some boilers should explode, out of the vast number in daily use on land and sea, in the factory and on the rail; it would be strange indeed, if that curse of humanity, carelessness, was not felt in its magnitude; for, reason and theorize as we may, it is a well-settled fact in the minds of scientific and practical men, here and abroad, that to this cause most of the accidents with steam may be traced. It is carelessness that makes boilers on bad plans, of poor workmanship and material; it is carelessness which omits the thorough inspection which boilers should have every thirty days; it is carelessness which permits crown sheets and flues to be burnt from scarcity of water, and water-bottoms, legs, and fire-boxes to be bent, burnt, and distorted from deposits of mud, scale, or refuse that is suffered to accumulate; it is carelessness which allows safety-valves to be jammed or overlooked, feed pumps to look after themselves, braces to be slack where they should be taut, and the pins in the braces not turned, or bent over, so that they cannot slip out; such cases have been known. It is more than carelessness which allows imperfectly welded wrought-iron sleeves for the socket bolts to be used to cover the same, for the water has free access through the open seams, and destroys the bolt as quickly as if there was no "protection." Cast-iron sleeves are now used in the best shops and besides being a perfect protection to the socket-bolts, they are more durable and much cheaper. From the first hours of its practical operation until the day of its final condemnation, a boiler is constantly growing weaker, and it should be so cared for that the work it is obliged to do is proportionate to its strength each year. To ascertain what the strength is, we must test it, and this can be done in a simple, cheap, and expeditious manner by water and heat. If a boiler be filled *full* of water up to the very safety-valves and all apertures closed, when a fire is built in the furnace, the water will be expanded, and raise the valve, if the boiler is strong enough to withstand the strain, but if it is not, the weakest part will be shown, and sometimes sheets are torn out by this method. Steam is not generated from the water during this test, and if a rupture does take place in the boiler no one will be injured by it. The safety-valve must be loaded to the utmost limit of strain that it is supposed the boiler will bear; and if the test is favorable, only three fourths of the load on the safety-valve must be employed for the working pressure.

It has never been proved beyond question that a steam-boiler exploded from any of the theories put forth in each disaster. Some persons have a passion for "explaining" matters that they do not understand by something else they are ignorant of; and we have had hydrogen gas brought forward as an agent in causing explosions; water suddenly flashed into steam as another; electricity for another; and so on, through the category. These are simply excuses on the part of some one at fault for the disaster. *After* a boiler has exploded, it seems almost supererogatory to go and look at it, and say what caused the disaster. We have heaps of smoking ruins, iron bent and blackened, and in most cases each part is a fac-simile of every other explosion; the torn sheets are gravely examined, and the conclusion arrived at is that "somebody was to blame."

We have no desire to treat the matter with levity, but is it not time that we had more careful superintendents of steam-boilers, and fewer inquests? In some cases, the cause of the accident may be pointed out after the explosion, but in such it might have been done equally well before. As we have before remarked, it is to be expected that some boilers will explode in spite of all inspection, just as cannon do with the most careful gunners, but it is a part, and a most important part of an engineer's duty to be thoroughly convinced of the soundness and strength of his boiler. When we see how seldom accidents of this kind occur to marine boilers we have positive proof of the value of thorough oversight and watchfulness.

## IS YOUR BOILER SAFE?

Every manufacturer, and every corporate body employing steam as a motive power, is immediately and directly interested in this question. In cities, particularly in the business part, the use of steam power is very general, and almost every square rod has its separate boiler. In view of this fact it behooves the merchants, capitalists, mechanics, and engineers owning or in charge of these boilers to see that they are always in as good order as they can be put. Experience has demonstrated the fact that very many boilers are not only out of repair, but totally unfit to be used at the pressure at which they are commonly worked; and not only is this a fact, but it is also true that these boilers have become so by carelessness and neglect. Without going further into this part of the question, let us see what can be done to avert the evil. Boiler explosions are continually occurring, and, so far from being solved as to the cause of them, remain as inscrutable as ever. While we cannot say, in every case, what the origin of disaster has been, we may at least avert the possibility of danger by paying some attention to the primary principles of boiler preservation. In the first instance, steam-boilers are too often neglected. Many engineers know little as to how the boilers are constructed or internally arranged, and of course they are incapable. If manufacturers will continue to employ such persons, when there is plenty of skilled labor in the market, they and the public must abide by the consequences.

Steam is generated from water; it is pumped in for that purpose. Too much water wastes coal, too little burns the boiler; the golden mean should be observed in all cases. The height of water in the boiler, measured by the gage-cocks, depends upon their distance from the crown sheet. In general the boiler-makers insert the first one at from four to six inches over the crown sheet of the furnace, the second one six inches above the first, and the third one at a like distance; thus, water issuing from the third gage-cock indicates that there is eighteen inches of water on the crown sheet. This amount is never, or at least should not be, carried in the boiler, as it is useless and wasteful, provided the tubes or flues are in the proper place. Keep the water, as a general rule, between the first and second gages, or as engineers call it, "scant two," and the best results will be obtained. On trying the water do not pull the handle of the cock round with a jerk; that may indicate a contempt of the force of steam, and be very knowing, but it is very bad practice. Open the cock gently and partially, so that only a slight aperture will be presented for the rush of the steam and water, and the actual amount of water in the boiler will be indicated. Steam naturally seeks an outlet; when the gage-cock is opened suddenly the steam raises the water to the opening, and does not, therefore, give a true exhibit of the water line.

There are, however, other things equally as important in the management of steam-boilers as the quantity of water, and these we will consider. Let us examine the braces and their relation to the work required of them. Take off the man-hole plate, and get in the boiler. It is a good plan to do this once in a lifetime at least, as we can then fix the general appearance in our minds. Take a flat chisel or a small iron bar and sound the braces; possibly some will ring like a bell, while others vibrate slowly, like a loosened cord; take out the latter and shorten them, they are too slack. See to the jaws of all the braces that they be not split or cracked; take out all those braces which are worn thin by rust or the action of scale or deposit; in a word, set them up to their work. In some boilers there are not braces enough. The crown sheet, particularly, is weak when it is flat, and requires the most consideration. Ascertain the pressure of steam on your boiler, divide the area of the crown sheet into inches, multiply the inches by the pressure, and you have the weight in pounds which the furnace top has to sustain. Reduce the pounds to tons and you will have some adequate conception of the crushing force on a fire-box roof.

Look also if there be scale forming on the flues

or crown sheet; if there is, it must be removed. Salt water deposits scale very fast (we have some in our possession over two inches thick, being a part of the lining formed in the boiler of the steamship *Cahawba* some four or five years since), fresh water not so quickly: in all hard water, however, there is more or less lime, which acts injuriously by forming a base to which the other salts and minerals, held in suspension by the water, adhere. Some engineers remove the scale by a pick hammer made for the purpose; they are not very efficient and are dangerous tools in the hands of inexperienced persons. Another plan to remove scale is to produce a sudden heat by a train of shavings or turpentine in the flues and fire-box when the boiler is cold; the expansion of the metal cracks the parasitic coating, and is supposed to remove it in sheets of greater or less size. This method works tolerably well, but cannot be relied on in all cases; it is also dangerous as being liable to burn the boiler. The best way to remove scale without harm to the boiler is to take either slippery elm, flaxseed, or any other mucilaginous seed or bark, and throw it into the boiler; by some peculiar action, which we are unable to solve, the scale is loosened and comes away in large pieces, and may then be removed through the hand-holes. We have tried the slippery elm with success in many cases, and it has never failed us.

If the stay and socket bolts of the water-spaces leak, they must be taken out and replaced; they are beyond remedy. The flues, by which the heat is arrested in its passage and imparted to the water, should be as clean as possible, and ought to be swept every week. Dust and cinders accumulated in them check their conducting power, and the temperature in the smoke-box will be much higher than it ought to be if they are neglected. So also with the deposits in the fire-box from the grate. Many ignorant engineers permit a stratum of ashes to lay around the edges of their grate bars and the water bottoms, supposing that leakage is prevented thereby. We desire to inform such slovenly persons that the leaks have been caused in the first place by this very practice; doubtless they apply the poultice on the principle that "like cures like." At all events it is a nuisance that should be stopped. So also with the water bottoms and legs of the boiler, keep ashes and water away from them and they will last longer. If leaks occur, stop them in a legitimate way; send for a boiler-maker and have them caulked, though an engineer should be capable of doing such work himself. It has been asserted that horse dung, slippery elm, flock, and other fibrous or flocculent matter would stop leakage in a steam-boiler. This may be a convenient practice, but it is an empirical one; a leak in a boiler is evidence of weakness in that part, the object is not to cover it up, but to strengthen it, and this can only be done by caulking or by renewing the seam with another sheet.

Finally, in reviewing this subject, let us assert that although the origin of steam-boiler explosions cannot be positively stated, we know that certain causes produce certain effects, and that neglect and carelessness have no business anywhere in mechanical matters, much less ought they be visible about steam generators. Boilers should be clean inside and out, and strong as well. It is of no use to put on dabs of putty to hide leaks, or to fill a boiler half full of horse manure for the same purpose. Make a radical cure, take out the bad part and replace it with a new one; if the cocks leak—the blow-off or gage—grind them in and make them tight. The blow-cock too often runs away with fuel that no one thinks of; it permits the heated water to dribble out, little by little; this water has to be supplied by other, less hot, thus incurring a needless expenditure. Stop it. Grind the plug in tight. Take care of the safety-valve, and don't have it sticking fast, or corroding, or leaking. Try it every day; load it properly and it will do good service, but not otherwise. How many engineers are there in this city or in others that will conform to the above rules—reliable and standard ones, and proved by actual practice and experiment?

## FAULTY CONSTRUCTION OF STEAM-BOILERS.

It is palpable to the close professional observer of the manner in which steam-boilers are generally constructed, that there is not only great need of reform in the actual workmanship, but that a large proportion of the accidents arising from the use of steam can be traced directly to the faulty construction. It is a truism that "the strength of any structure is exactly that of the weakest part;" but who can say where the weakest part of a steam boiler *is*, as they are ordinarily made? Take a simple cylinder boiler, for instance: the sheets are run through the rolls and bent to the proper radius; when the riveting gang get to work, they close up the rivets with great rapidity, but when the holes come out of line with each other, the drift pin is restored to, and the sheets are literally stretched until the rivets can be inserted; when the drift pin is knocked out, the sheet goes back to its place, and there is already, when tested, strain enough to weaken the boiler. Repeat this performance through twenty or thirty feet, the length of an ordinary cylinder boiler, and who can say where the weakest point of the structure is? Suppose such a boiler to be made of silk, for instance, or any flexible material, what shape would it be in? It would be full of puckers, folds, seams, and gathers, and represent most accurately the various trials to which that most abused of all modern engineering apparatus—the boiler—is exposed.

The case is aggravated, not benefited, when we construct a square boiler, for this shape seems, by general consent, to have been adopted for marine service. When the angles or flanges of the sheets are not broken by the flange turners, they are cracked out by the drift pin of the riveting gang, and it ought to be made a capital offence to have such a tool on the premises of any boiler-works. New boilers burst under the most mysterious circumstances; old boilers are patched and then burst; and we are told gravely that "putting new cloth into old garments" is the solution of the trouble. On each occasion the Coroner examines a host of "experts," who proceed to declare that "the iron was burnt," "the water low," "the stays insufficient," "the water changed into explosive gases," etc.; but it never occurs to these worthies that the actual strength of the boilers was in many cases unknown, and that the boilers may have been at the bursting point for days, weeks, or months, until at length they gave way.

It may be argued against this view of the matter that, if hydrostatic pressure is applied, it makes no difference where the strain comes, for the boiler is, as we have admitted, just as strong as the weakest point. It must be borne in mind, however, that it is natural or only reasonable to infer, in theory at all events, that every square inch of the boiler sustains an equal strain; with faulty construction this is impossible, for there may be, as we have shown, almost a rending force without a pound of steam in the boiler. It is ridiculous to suppose that safety is secured by extra heavy iron; the best materials and the finest workmanship in other respects are of no use so long as rivet-holes shut past each other, so much that some rivets we once took from a boiler were offset nearly half their diameters. Holes will come out of truth with the most utmost care, especially in such hap-hazard work as punching is generally made; and when they do so, the only safe way is to rivet all the true holes first, rim all the faulty ones to one size, and then put rivets in that fit, just as a machinist turns bolts to fit true holes in a bed-plate or cylinder. This method is no doubt costly, and will never be adopted, but it has the merit of common sense if no other. There is a great deal of carelessness in caulking seams also; for when the chipper chamfers the edge of the plate, the lower side of his chisel bears on the sheet and leaves a furrow; not very deep, it is true, but sufficient to cut through the skin of the iron, which is the strongest part. Neither are the braces properly set, for some draw all one way while others don't draw or hold at all, and are perfectly loose; thus a portion do all the work, and the rest are idle. They impart no strength and are an element of weakness, for the engineer relies upon them when they are doing no good. We are confident that a great deal of attention can profitably be given to the mere workmanship of steam-boilers; they are not tanks or receptacles for boiling water, but great magazines wherein a tremendous power is stored, the safe custody of which is of paramount importance to all in the vicinity.

---

## TROUBLES INCIDENT TO STEAM-BOILERS.

"I don't see what is the matter with my boiler," said a friend recently; "it used to make steam enough, but now it is all I can do to run the engine through the day." Upon having an examination, the mystery was found to consist of ashes in the smoke-box, and soot in the tubes. Simple enough, certainly. The cure was a shovel and half an hour's labor.

Many people have an idea, apparently, that a steam-engine loses some portion of its vitality every year in some unknown way, so that its decline and fall is simply a question of time. This is true where no care is taken of machinery, but, with intelligent supervision, and repairs when needed, a steam-engine one hundred years old will be as good as the first day it took steam. It is as unreasonable to expect a steam-engine to run continually without repair and inspection, as for a human being to exist without eating. A little reflection would show that if a steam-engine has run for a term of years, doing the same work continually, the failure, if there be any, arises from natural causes, and that examination of it by a competent person would be the course to adopt.

It often happens that shafting gets out of line in a shop, and that the machines generally are disordered in their relation with the power which drives them. Where this is the case, lining up the shafting and setting up the machines again would effect a great saving of power and fuel. It also happens that boilers sometimes give out, or cease to make steam freely, from the destruction of the draught.

If one building be erected by the side of another, the draught of the chimney will be affected when the wind is in a certain direction, and this in spite of the general cleanliness and good condition of the boiler. The remedy for this is to increase the height of the chimney or put in artificial draught.

It is also frequently the case where pine wood or bituminous coal is used for fuel that a resinous deposit forms on the inside of the tubes, to the very great detriment of the steaming qualities of the boilers. It is extremely difficult to remove this, as it is composed of soot and resin, and adheres to the iron with great tenacity. A whalebone brush is sometimes employed; also a brush made of steel wire, but these instruments merely scratch the surface of the deposit without removing it. It has occurred to us that a strong, hot solution of potash might be used with good effect in this case, and we recommend a trial of it at least. It cannot hurt the boiler externally, and is so easily tried that it should be.

Another acquaintance, some time since, called our attention to his boiler and engine, the boiler failing to make steam sufficiently, although in size it was ample. The defect here was in the setting. The boiler, an ordinary cylinder, was set on top of two brick walls, as the cover would be laid on a box, and the fireplace was simply a gaping cavern, in the further end of which the throat of the chimney loomed wide and voracious. If all the heat of Vesuvius in eruption were turned under the boiler it would hardly make steam enough in its condition. The steam would have been made in the chimney, for that was where the heat went, and its effect on the boiler seemed more like a passing favor than any actual duty it was bound to perform. When the furnace doors were opened a roaring wind passed through them, and the blaze went far up the chimney. The remedy in this case was to lessen or obstruct the draught; to add a bridge wall five or six feet from the furnace door, and to put a damper in the chimney, so as to arrest the heat when desired.

The field for improvement is very wide. The proportion of heat utilized to that driven off or lost is very little—hardly one tenth—and this waste is going on continually. Of course the quantity differs in different boilers, and can be greatly lessened by good management, but that great slovenliness in the use of fuel, and great indifference prevails on the part of proprietors toward getting competent engineers to attend their boilers, is apparent to any intelligent observer.

---

## STARTING FIRES UNDER BOILERS.

A very mischievous practice exists in various parts of the country, in reference to starting fires under steam-boilers preparatory to raising steam; this duty is entrusted to ignorant watchmen, who are too often the agents of disaster. Those men are instructed to light the fire at a certain hour, and generally comply with their orders without exercising the least judgment on the subject; they rarely try the gages to see that there is water in the boiler before fulfilling their duty. We can call to mind several accidents or injuries that have occurred to boilers from this very cause. The Detroit Locomotive Works once had a boiler heated so hot, by the carelessness of a watchman, as to burn the felt lagging on the outside; and many other similar cases might also be cited. We have known instances where watchmen have started the fires under gangs of cylinder boilers, and raised steam in them to such an extent as to drive the water out of some into the others not in use, or not so full; thus running the risk of burning the boilers and causing no end of delay and loss. The men in question ought not to be permitted to meddle in any way with a steam-boiler, and no persons except those who are skilled in the management of them, and who are conversant with the properties of steam, should under any circumstances be entrusted with their control. Too many lives have been lost and too much property scattered to the winds by the ignorance of those who were temporarily left in charge of boilers.

---

## STEAM-BOILERS AND ELECTRICITY.

Boilers have burst under every possible circumstance and in every condition—while the engines which they have driven were at work and while they were quiescent—with low steam and high steam—with water and without water, and under mysterious circumstances apparently the most impenetrable. Formerly it was a generally received opinion that the contact of comparatively cold water with an overheated plate, generated an excessive amount of vapor of an especially dangerous character, the expansive force of which no form of boiler nor any diameter of safety-valve could operate against effectually. So generally was this opinion received, that all explosions were at one time attributed to it, and the engineer who was so fortunate as to survive his disaster, was universally discredited when he asserted that there was plenty of water at the time of the accident.

But lo! certain inquisitive men—and it is to them that science owes all her discoveries—quietly take a boiler, heat it to redness, and then inject water in quantities. So far from blowing it up, the vapor only discharges itself through the safety valve with a mighty roar!

This theory, as an universal and general source of danger, has gone to the clouds with the puffs of steam that destroyed its value. Perhaps the latest cause assigned as the mischievous force which destroys steam-boilers by explosion is that of electricity. We find an account of an apparatus once used to ascertain the presence of this agent, and the manner of its generation in steam, in a philosophical work:—

"The apparatus was a common high-pressure steam-boiler, about three feet long and twenty inches in diameter, mounted on insulating pillars, and strong enough for a pressure of 200 pounds to the inch. The steam was suffered to escape by jets of a peculiar form, on the side of a box into which it was admitted by a cock. Faraday, in investigating the electricity of steam, found that dry steam gave no excitement, and that the electricity resulted from the friction of vesicles of water against the sides of the orifice. Hence the box contained a little water, over which the steam escaped, and was partially condensed. The jet had an interrupted passage to produce friction, and its nozzle was lined with dry box or partridge wood. The vapor escaped against a plate covered with metallic points, to collect the electricity, and ending in a brass ball insulated from the earth. The boiler was negative, and positive electricity was collected at the ball, provided the water was pure and free from grease. Turpentine, and other volatile essences reverse the polarity, while grease or steam from acid or saline water destroys all excitement. If the nozzle of the jet ends in ivory or metal, there is also no excitement. A boiler, such as is described, will develop in a given time as much electricity as four plate machines forty inches in diameter, making sixty turns a minute—a truly surprising result.

Thus it appears from high authority that electricity can only be obtained in steam under extraordinary circumstances. Certain features in the detective apparatus must be rigidly conformed to, otherwise it fails to appear. And what is sufficient to utterly nullify any value this theory may have had, is the fact that the presence of grease or steam from salt water prevents the electrical fluid from manifesting itself. As steam-boilers are rarely, if ever, free from oil in small quantities, it will be seen that there need be but little danger apprehended from boiler explosions through electricity.

## FIELD FOR IMPROVEMENT IN STEAM-BOILERS.

The boiler of a steam-engine costs more than the engine; and, considering the wide use and valuable service of this prime motor, there is, perhaps, with the single exception of the plough, no instrument of more importance. There is, perhaps, also, notwithstanding all the inventive faculty and experiment that have been expended upon it, no instrument more imperfect. A boiler of ideal perfection should secure complete combustion of the fuel, so as to obtain all the heat which the coal will yield; it should transfer this heat to the water to form steam, and it should hold the steam in absolute security. In practice, very few boilers effect complete combustion of the fuel, and none secure the transfer of nearly all the heat to the water.

When anthracite coal is the fuel used, the only portion of it which is of any value is its carbon. The burning is the combining of this carbon with the oxygen of the atmosphere. Carbon combines with oxygen in two proportions—one atom of carbon combining with one atom of oxygen to form carbonic oxide, and one atom of carbon combining with two of oxygen to form carbonic acid. According to the experiments of Favre and Silberman, one pound of carbon, burned to carbonic oxide, will raise the temperature one degree of

Fahrenheit's scale, of 4,451 lbs. of water, while a pound of carbon, completely burned to carbonic acid, will heat one degree Fahrenheit 14,544 lbs. of water. Hence, coal burned only to carbonic oxide generates less than one third of the heat of which it is capable. When coal is burned with an insufficient supply of air, either the whole or a portion of the product of combustion is carbonic oxide.

But the greatest loss of heat in steam-boilers is the failure to secure the transfer of all the heat generated from the products of combustion to the water. In order to effect this as nearly as possible the tubes should have the thinnest walls practicable, as we recently pointed out. It is also quite as important that the walls of the fire-box should be of thin plate.

Heat is radiated from all substances with a rapidity proportioned to their temperature. When, therefore, two bodies of different temperatures are placed in contiguity, the warmer will send its heat into the other more rapidly than it will receive heat from the other in return, consequently the cooler will be warmed with a rapidity proportioned to the difference in the temperatures of the two. The same law applies to the transfer of heat by conduction from one body to another; it takes place with a rapidity in proportion to the difference of the temperatures.

Suppose we have a fire-box plate four inches in

thickness, with fire on one side and water on the other; the surface next the fire may be red-hot, while that next the water is only 250 or 300 degrees. There being, then, but little difference between the temperature of the gaseous products of combustion and that of the contiguous surface of iron, the transfer of heat from one to the other goes on slowly; and the same is the case with the transfer of heat from the interior surface to the water. In this case the products of combustion go up the chimney at a high temperature, carrying away nearly all the heat generated. If the plate is thin there can never be this great difference in the temperature of the two surfaces—the surface next the fire will be cooler and that next the water will be hotter; the transfer of heat will therefore be more rapid, and the rapidity will be in proportion to the thinness of the plate.

The transfer will also be proportioned to the rapidity of the circulation. Water is one of the poorest conductors of heat, and if a stratum next the plate remains in its position, so soon as it is heated to the temperature of the plate the transfer of heat ceases, or goes on with the slowness with which heat is conducted away by the water; but if, the instant a particle of water is heated, it is replaced by the coldest one in the boiler, the transfer of heat goes on with the greatest possible rapidity. In plain kettles, heated from the bottom, the ebullition creates a very active circulation; but in small tubes, if the bubbles of steam are passing in one direction and the water in the opposite, the circulation is seriously impeded. Inclining the tubes, as in Dickerson's boiler, is an exceedingly simple and effectual means for producing the most active circulation, and is probably destined to be very generally adopted.

The most effectual plan, however, for insuring the transfer of all the surplus heat is to pass the products of combustion right into the water. This plan has been tried on a steamboat on the North River—the *John Faron*—but was abandoned in consequence of the accumulation of ashes in the boiler; it would seem, however, that this difficulty, being merely mechanical, ought to be overcome, in view of the great advantages to be realized. It is true that both the air and the fuel would require to be introduced against the pressure of the steam, but as the air would be worked through the cylinder, its expansive force would doubtless be sufficient to drive the air pump. Prof. Seely has suggested that the carbonic acid might be absorbed by the steam, so that no increased tension would result from it; but this certainly would not be the case with the nitrogen, and its expansion would at least prevent any loss of power. This plan would give not only the most effective and economical, but also the simplest and cheapest, of all conceivable boilers. All that would be required would be a plain cylinder with an inclosed fire-box, without any tubes, stays, or other costly adjuncts; and, as no heat would pass through the shell, it might be of any thickness necessary to insure absolute safety from explosion.

All that is required to make this great improvement practicable is some simple and effectual plan for preventing ashes from going into the boiler, or for readily blowing them out after they are introduced.

---

## CHAPTER XXXVII.

### LOCATION OF STEAM GAGES AND INDICATORS—THE LAWS OF EXPANSION.

A FRIEND informed us recently that he had two steam-boilers connected by a pipe which is furnished with a stop valve for closing the communication between the boilers. He had the valve closed, and found that the pressure in one boiler was fifty pounds to the square inch and in the other twenty. On opening the valve the pressure immediately rose to sixty-five pounds. It would be interesting to have further particulars in regard to this experiment, but with our present light we are inclined to attribute the surprising result to the location of the gage in such position that it was acted upon by the current of steam in its passage from the high-pressure boiler to the lower.

The action of currents of steam, though familiar to engineers in other situations, seems to have been strangely overlooked in its effect upon gages and indicators. Clark, in his most able work on the locomotive, states that repeated observations showed the pressure to be greater in the steam chest than in the boiler, and he remarks that from the carefulness with which the observation was made, and the perfection of the instruments, it is as difficult to doubt the statement as it is to believe it. There may be difficulty in doubting the statement, but to believe it is simply impossible. Steam will not flow from a vessel of lower pressure into a vessel of higher pressure. There must have been some error in the observation, and a very probable cause of this was the location of the gage in such position that it received the impact of the swiftly moving current of steam which rushes from the boiler into the steam chest.

Currents of steam may operate not only to raise the mercury in a gage, but also to lower it so as to indicate no pressure whatever, even in engines working steam at a pressure of thirty or forty pounds to the inch. This effect is produced by inserting the gage pipe at right angles to the current of steam, when the steam is drawn out of the pipe by the friction of the passing current, and we may even have the indication of a partial vacuum. This matter is worthy of attention on the part of the builders and runners of steam-engines.

---

## THE LAWS OF EXPANSION.

Since the steam-boilers are greatly affected by constant contraction and expansion, this article will be appropriate here.

The constituents of heat have long remained a problem for scientific men; but up to the present time no satisfactory solution of its peculiar nature has been accepted by the world at large, although very many ingenious theories have been promulgated respecting it. While this is true of heat itself, its action upon matter—solid and liquid—is, in some cases, well understood, and the laws relating to expansion are clearly defined.

The general fact that most bodies expand under heat is well known; the proportion and nature of the change, however, which takes place in the object heated, is not alike in every case. Different solids or substances of dissimilar nature expand unequally, but as a rule, uniformly, and resume their former shapes and dimensions when cooled. This statement, however, must be qualified by the remark that there are exceptions in the case of certain metals, as iron, steel, and brass. While the assertion made above is theoretically correct, the fact is unimpeachable that the metals specified do enlarge permanently in bulk with successive heatings, so that it is perfectly possible to remedy spoiled, or damaged work in the machine-shop by this method—that is reheating. To illustrate this in a practical manner, take a crank pin, for instance, that has been turned too small in the conical end, so that while it fills the hole in the crank it does not fit it; let this pin be heated and cooled in water three times (not so that scale will form), and it will be found that the metal has gained in size, having absorbed some element that caused its fibres to swell and so enlarge the diameter. Ligneous or woody substances also expand more sideways than lengthwise, and, when greatly heated, *contract* permanently and remain fixed. Argillaceous or clayey substances, such as pottery, contract by heat; in this case chemical changes take place which alter the nature of the material. Lead is an utter exception to the general law of expansion, as, when under the influence of heat, the particles of metal slide over each other, and do not return to their former shape when cooled. Lead pipes, when used to convey hot water, become permanently elongated, as may be seen by examining those that have been in use for years; in most cases the fastenings will be found loosened and the pipes distorted. Bath-tubs and other vessels lined with lead have the same shrivelled or wrinkled-up appearance, showing that the metal has undergone alteration in form since it was first applied.

The amount of expansion in solids between the extremes of zero and the boiling point of water (212°) is comparatively little; zinc, one of the most easily affected by heat, elongating but 1-340th of its length, glass expands only about one third of this quantity during a similar heat. The following list exhibits the ratio of expansion between different metals in the order in which they are named: zinc, lead, tin, silver, brass, gold, copper, bismuth, iron, steel, antimony, platinum, and glass. This is also very nearly the order of the compressibility of metals.

When expansion is uncompensated for in machinery, a tremendous disturbance takes place, often causing it to cease its functions entirely; steam pipes are torn and twisted from their fastenings; bed-plates broken, and shafts bent by this uncontrollable force. An iron rod, one square inch in section, when raised from 32° to 212° expands with a force of 35,847 pounds, or it exerts a force of 199·15 pounds for every degree (Fahrenheit) that the temperature is increased. Some phenomena observed in daily life may be traced to the laws of expansion, as, for instance, spikes driven into wood gradually enlarge the holes and loosen themselves by the changes in temperature they undergo. Iron and platinum wires may be cemented firmly to glass without danger of breaking, because they expand in nearly the same ratio; but gold, silver, or copper cannot, because their degree of expansion varies from that of glass. So also railroad tracks must be laid with a space between the rail ends, otherwise the whole line would be disturbed. Accidents have frequently occurred from this cause. The same features are also observed in iron bridges. Time-measures suffer much from unequal expansion, as where it is uncompensated for, a great change is observed in the record. Almost every material thing on the globe is affected by expansion, or the influence of heat, at some period or another; and yet the physics of this mighty agent are still undiscovered.

# PART V.

# GEARS, BELTING, AND MISCELLANEOUS PRACTICAL INFORMATION.

## CHAPTER XXXVIII.

### RELATING TO GEARS.

IT is very apparent to the experienced observer that many—may we say *most* of the gears in use?—have been designed with very little reference to the laws which govern their motions—laws which are as fixed and as immutable as the orbits of the planets. We refer in this connection to cut gears. Both kinds have their spheres of action—the cut wheel on light and delicate machinery, and the cast wheel on heavy work; which, in general, from the great size and weight of the parts, preclude the possibility of cutting them except by chipping, which injures instead of benefiting them.

Every wheel that comes from the foundery has a vitreous scale on it, caused by the partial fusion of the sand and facings, which makes it wear well and run easy, and it seems the acme of folly to remove this and expose the softer part below to the abrasion experienced at work.

What is it that makes wheels rattle and jar at work when all the details are firm and solid? And what is it that springs shafting, breaks hangers, forces them out of line, and keeps bearings chronically hot so that all the oil which can be poured on is of no avail? It is teeth grinding one on the other; it is wheels with teeth out of time; it is wheels with teeth of fancy shapes, ill proportioned and entirely defective as regards their essential points. There is but one correct shape for a tooth. Man may scheme and plan as he will, to one shape it comes at the last. Nothing within the whole range of mechanical movements is more beautiful and simple than the transmission of power from one wheel to another by two projections rolling on each other—rolling but never rubbing; each point passing the other with no more than rolling friction, which is the least of all resistances.

That wheels shall so act, depends entirely on the shape of their teeth, supposing, for brevity, that all other mechanical parts are right. If we make teeth pyramidal in form, as they are made in one of the largest and oldest machine shops in New York, they grind and rub for years until they have worn a path for themselves that they can follow with ease. Then tooth and space are alike out of proportion, and with any change of velocity, backlash ensues. One lags until the other catches up to it, and rattle, bang, and clash is the order of the day. If we make teeth semicircular at the top and semicircular at the bottom of the space, we take away part of the impelling surface in order to make the wheels mesh properly, and we load the apex or crown of the tooth with a portion of metal which exerts no useful effect but entails waste of material and of power.

The proper form of a tooth is the epicycloidal, or that curve which is generated by one circle rolling on another circle. As this curve is never constant, but changes with every diameter and pitch, all the circumstances or conditions must be taken into account before wheels which will work mathematically correct can be made.

If this well-known fact was rigidly followed out, however, the greatest confusion would ensue, and for every wheel needed, a different set of cutters would have to be made. This would not only be costly but wholly impracticable in the ordinary business of the workshop. A manufacturer who wanted a gear cut without delay, would hardly be disposed to wait until the foreman could go to his factory, look at the size of the driver, count the teeth, make templets and then cutters before his order could be filled. The variation in the shape of a curve, therefore, is so small that it need not be considered in practice, and for wheels of all sizes, such as change wheels for screw cutting engine lathes, a few cutters will answer, commencing, we will say, at those for 50 teeth and running both ways to 12, which is about the lowest gear.

A common way of finding the shape of teeth is to strike the curve from the pitch circle of the next tooth. Not the centre of the tooth, *but its side.* This and a subsequent process gives a curve approximating to an epicycloid; not a true one by any means, but a simple and ready way of forming a tooth that will run and wear well, and a method that is within the reach of every one. The curve so struck from the pitch line outward to the crown of the tooth gives the working side, while lines drawn from the pitch line to the centre, give the base of the tooth. The result of these will give an angle more or less acute at the pitch line governed by the size of the wheel, which is of course to be rounded off.

The only correct way to lay off the teeth is on the pitch line, not on the chord of the pitch line. The way to arrive at this is to step off, by an infinite number of small spaces, the pitch on the arc, and then to take the chords found as a constant by which all the other distances may be obtained. It is usual for the pitches of pinions of small diameter, working in large wheels, to be a little greater than the wheels they drive or are driven by.

A true epicycloidal curve may be formed by

making segments of wood of the size of the intended wheel, and gluing thin slips on the back of them; something like the flanges which are often made on pinions. The curve of the wooden segment represents the pitch line of the wheel, while the thinner slip projecting beyond it represents the crown of the tooth or outside diameter. Two small steel wires—stout needles will answer—must be pushed through the slips, so that the points come out exactly at the pitch line and at the crown. By placing these segments together and rolling them over each other, as the wheels act when at work, true epicycloids will be formed on the inside of the slips, from which patterns can be taken to shape the teeth by. The points must not project too far through the wood, or they will stick in and not mark. It will be found that the shape thus given is acute at the point of the tooth—too much so for common use, and it must therefore be cut off, but nothing must be added to or taken from the working curve.

It is not at all difficult to make a set of wheels revolve one another. If we take a blank, put it in a milling machine or an arbor, set the index, and start the cutters, a wheel will be produced that will mesh when two are put together, but something more than meshing is needed, and when the teeth are not correctly laid off—or rather shaped—one face will slide on the other face instead of rolling over it.

Beside the shape of the teeth there are other points essential to be considered. These are the width and depth in proportion to the pitch.

The best English practice makes the width of the tooth of spur gears for ordinary work, or where the velocity does not exceed five feet per second—an axial width equal to four times the thickness, and five times the thickness where the velocity exceeds five feet per second. The length is given at five eighths of the pitch, and never to exceed three fourths of the same. Outside of the pitch line the length is four twelfths, below it five twelfths; which, in the neat length of the whole, from bottom of space to crown of tooth, equals three fourths of the pitch.

In an absurd hand-book recently published we are told that the pitch line is the centre of the tooth. This work purports also to give directions how to make and cut gears. No set of gears can run properly in that way. The working diameters are at the pitch lines, and the centre of motion is also there in properly made wheels; but if the middle of the tooth is the pitch line, the gears will bottom when at work, or else the pitches cannot coincide. This is the way tons of gearing are made in this country, and it is needless to tell men of common sense that it is wrong. Wheels must have good clearance at the bottom, and have their pitch lines coincident at all times to obtain the best result.

Quantities of cast gears are used on reapers and mowing machines, but they run badly. Sometimes inaccuracies occur through shrinkage, and in others, patterns which were once perfect are destroyed by time and ill usage so that they are no longer reliable. All such should be looked after, condemned if beyond repair, and new sets made on proper principles.

No set of cast gears can run quietly and perform well otherwise, when they are bored out untrue; where they are keyed on so as to be set to one side, even if they are true, or where they are oval through imperfect shrinkage, as before remarked. There is a draft side on every pattern, and cast gears should be meshed so that the draft sides are opposite each other, but in more than one instance we have seen men carefully filing the teeth down "because they were thicker on one side than the other," thereby wasting time and spoiling files.

---

## CHAPTER XXXIX.

### LEATHER BANDS—BELTING.

THE horse-power of belting or the tractive force exerted by leather bands of a given width, at a certain speed expressed in foot-pounds, or in any other positive way, is not generally known. We do not know what it is, although we have some half dozen rules professing to give a unit for a horse-power, which are obviously incorrect. A horizontal belt of a given length will drive more than a vertical belt of a given length; a long belt more than a short one, and a twisted belt more than either, because in the case of the horizontal and the long belt, the sag and weight tend to produce closer contact and resist strain better than where the belt merely hugs the pulley by its tension; the same is true of the crossed belt, which embraces more of the circumference of the wheel driven.

Eight hundred feet per minute velocity for a one inch belt is said to give a horse-power; four hundred for a two inch belt will give the same; and soon in reverse ratio.

The dynamometer affords an easy, simple, and cheap method of testing strains, or the transmission of power from one machine to another, and a few experiments by it would settle forever all doubts and uncertainty on this point. The dynamometer merely weighs strain as a butcher weighs meat, and with the same instrument—a spring balance. If a lever be made with a bearing, cap, and bolts at one end, and the same fitted to a shaft, and if a spring balance be applied to the other, by weighting the lever until it balances the tendency to raise imparted to it by the shaft, we shall have an exact record of the actual number of foot-pounds of work or strain exerted by the machine tested, when the relations between the diameter of the shaft and the length of the lever are considered. Of course, with such a dynamometer there is great friction, and if the test is continued long, much heating on the shaft occurs, which would interfere with a correct result; one sufficiently correct for practical purposes may, however, be obtained if the experiment be properly made.

There are many other forms of dynamometers for weighing or observing the force of machines, but it seems unnecessary to consume space with details of them, when it is apparent to all persons, who would be likely to undertake the experiments here recommended, what such apparatus should be.

Some things relating to the action of belts are but imperfectly understood, for although Morin's experiments have demonstrated the relative resistances of belts on pulleys of different materials and surfaces, such as rough cast-iron, smooth cast-iron, wood, etc., he has not informed us of their position, their nature, whether vertical, horizontal, or twisted, and whether the ratio of resistance increases in regular progression from a belt one inch in width at 400 feet up to a belt 30 inches wide, at the same velocity. It is obvious that these matters exercise a great influence on the transmission of power by belting.

From an experiment at one of our largest machine shops, it was found that gearing absorbs less power than belts, and that the force required to work the latter is extremely variable, depending upon the tension, the condition of the surface of the pulley, and minor matters. This fact was deduced from observing the working of a fan blower, and is to be received with caution, for it has hitherto been supposed that gears consumed more power than bands, and these results may be due merely to the peculiar arrangement of this special machine. It is a fact, however, that the use of sawdust, resin, or similar substances, to increase the adhesion of the belts to pulleys, as also the employment of idler pulleys, or rollers suspended against belts to keep them up to their work, as well as the divergence of belts from right lines or carrying them at acute angles about rollers fixed in walls, add greatly to the expense of working them.

---

## BELTING.

Certainly nothing can be of greater importance to manufacturers than belts, and all relating to them, for there is not a factory in the land, of any size, but has thousands of feet in daily use. Further, they are costly to replace, and careless or ignorant persons frequently destroy them by misuse.

Great remissness in lacing belts and laxity in matter of inspecting them frequently, to see if they need repair, is noticeable. We have seen large machine shops stopped for hours while the main belt was being laced, and it is nothing uncommon for half or three quarters of an hour to be wasted in stretching or putting in rivets, when the same ought to have been attended to over night, or, at least, during noon hour.

Manufacturers know very well that half an hour deducted from the labor of a machine amounts to a large sum, where there are many machines, and when these petty losses are easily avoided, there is certainly no excuse for their occurrence. Some man of experience should be paid extra to lace the belts whenever they need it. Let him make it his business to inspect them regularly, and be held accountable for their failure, if it appears that his neglect was the cause. This relates, of course, to the principal driving belts, for on the individual machines each workman ought to take care of his own.

The ends of a belt should always be cut off square, not guessed at by the eye, but laid off with a tool. The holes ought to be made with a small punch at a proper distance from the end—the size of the holes and the distances of them depending on the width of the belt. The use of an awl is reprehensible, for the holes are apt to be made irregular by it, and much larger than there is need of. The end of the lace should be tied with a square knot in the middle of the outside, for the corners of the belt where it is cut are most exposed and apt to whip out. Tying a belt lace does not look so neatly as where the ends are put through an incision, but tying saves the belt from having extra holes made in it. The laces ought to be of the same thickness from end to end, or as nearly so as possible. It oftens happens that laces have very thin spots in them; such should be kept for short belts, and never used for long ones. Moreover, the holes must be made at equal distance apart, and not too many of them; every hole weakens the belt, and none that are not absolutely essential should be cut. All new laces, as well as new belts, should be stretched by hanging weights on them before they are used—petroleum, sawdust, resin, and similar substances should never be used. When a belt gets harsh or dry, neat's-foot oil is the best thing to apply to it.

Since belts are so universally employed, a series of experiments on this subject would be invaluable. The power which belts are capable of transferring is so variously estimated by different persons, and sometimes so erroneously, that the only standard or conclusion that can be arrived at is that derived from actual practice. For what has been once done, under certain conditions, would be again. We here publish certain results observed by John A. Roebling, Esq., an engineer of note. They will serve as a guide in similar cases.

"Facts appear to be wanting concerning the

power of belts. Here are some, well observed and conclusive, so far as they go:—

"The towers of the Cincinnati bridge are 200 feet high, and contain 16,000 perches, or 400,000 cubic feet, of masonry each. This material has been raised by engines of 10 inches bore and 20 inches stroke, working with a pressure of 60 to 80 pounds, making 80 to 150 revolutions per minute. The power is transmitted by a 9 inch leather belt, from a 4 feet iron pulley, keyed upon the crank shaft to another 4 feet pulley, fixed upon a pinion shaft. This pinion is 14½ inches in diameter, and drives a spur wheel of 6 feet diameter; another 14½ inch pinion on the shaft of the latter moves a second spur of 6 feet, fixed upon the same shaft, which turns a 3 feet drum, which winds and unwinds a hoisting wire rope, of 1½ inches diameter. By this rope the weights are hoisted direct without any further tackle or appliances. By 25 revolutions of the engine the wire rope drum revolves once, and pays off or receives 10 feet of rope. This makes the speed of the belt 50 times the speed of the wire rope.

"A block of sandstone, measuring 60 cubic feet, and weighing 8,400 pounds, is a full sized stone for the work. The belt will run up each a block, the engine making 125 revolutions per minute, at the rate of 50 feet per minute, which task requires the tightening pulley to be pressed down hard, so that about ⅗ths of the circumference of the 4 feet pulley are closely hugged. The same belts have been performing this duty nearly three seasons without ever failing. On the contrary, a limestone of the same cubic contents, weighing 170 pounds per foot, or 10,200 pounds, cannot be raised without the slipping of the belt, and without such hard application of the tightening pulley as to endanger its splicings and safety. A weight of 8,000 pounds may be considered in this case as the fair working limit of the power of the belt. A strain caused by 10,000 pounds is altogether too much for safety and economy.

"The speed of engine being 125 revolutions per minute, the speed of belt is 4 × 3. 1416 × 125= 1571 feet per minute. The duty performed in this case is equivalent to 50 × 8,000=400,000 foot-pounds. With a load of 8,000 pounds the tension of the belt is 254⅝ pounds. Its speed being 1571 feet per minute, its performance is 1571 × 254⅝= 400,000 foot-pounds. Assuming the width of belt at 10 inches we have 40,000 foot-pounds for one inch of belt. The old rule allows one inch for every horse-power of 33,000 foot-pounds, and I think this is a very good rule for ordinary mill practice, provided the speed of belt is equal to about 1,500 pounds per minute.

"Blocks weighing 8,000 pounds have been frequently raised 150 feet high in 2½ minutes without the slipping of the belts. This speed is equal to 60 feet per minute, and the duty performed is equivalent to $60 \times 8,000=480,000$ foot-pounds= 15,54 horse-power; speed of belt=1,885 feet.

"The principal element which determines the power of a belt is its speed. A slow moving belt cannot transfer much power, any more than a slow moving piston. The higher the speed the more power will be run off. Now the question of speed can only be qualified by the question of wear and tear, and by adhesion. If the speed is too high the belt will slip, more pressure becomes necessary, and a greater wear and tear will result. Where there is no absolute necessity, the speed should never exceed 1,500 feet. A less speed of 1,000 to 1,200 is preferable and will be found more economical in the end. Where a higher speed, say 2,000 feet and more are essential, as in the driving of fans, economy must be neglected.

"The absolute strength of a belt is a fixed and invariable quantity at any one time. Speed, on the other hand, may vary, say from 500 feet per M. to 3,000 feet per M. The strength being given, the other factor (the velocity) will determine the performance, or foot-power which can be raised—provided in all cases that there is no loss by slipping. Now the adhesion of a belt is directly as the pressure. Ordinarily a belt or wire rope, passed over a pulley, will produce adhesion equal to its tension, resulting from a contact of a semicircle. In other words, we can elevate 2,000 pounds by a counterweight of 1,000 pounds, if the physical conditions are favorable and the speed is slow. As the circumference of contact is increased so is the adhesion. By taking a sufficient number of turns around a snubbing post, any line or rope may be broken, provided the post stands.

"The mathematical consideration of this question is very complicated, while the practical issue is easily determined by experiment. The adhesion of belts is also favored by large pulleys, but not in proportion to the size. The conditions of the belt, and also of the atmosphere, will likewise influence its performance.

"In conclusion I will observe that the engines mentioned above drive other hoisting gear, besides the drums. If the wire rope gear alone was to be operated, the power might be applied direct without any belting."

---

## CHAPTER XL.

### CONE PULLEYS FOR GIVEN VELOCITIES—FORMULÆ FOR CUTTING SCREW THREADS.

SUPPOSE the velocity of the upper or driving cone to be 100, the joint diameter of the two cones 20 inches, and the velocities required 75, 150, 225, 300. Write down the velocities required as above, and under each write that of the upper cone; add them together, and set the amounts under, as in addition, and make each amount the denominator of a fraction, of which the velocity of the upper or driving cone, say is the numerator. Multiply the joint diameter, 20, by each of the fractions so found, and the products will be the several diameters of the pulleys upon the driving cone. The same operation, repeated with the velocities sought, as denominators, will give the diameters of the driven cone. Example:—

| 75 | 150 | 225 | 300 |
|---|---|---|---|
| 100 | 100 | 100 | 100 |
| 175 | 250 | 325 | 400 |

$$100 \div 175 \times 20 = 11\tfrac{3}{7}$$
$$100 \div 250 \times 20 = 8$$
$$100 \div 325 \times 20 = 6\tfrac{2}{3}$$
$$100 \div 400 \times 20 = 5$$

$$75 \div 175 \times 20 = 8\tfrac{4}{7} + 11\tfrac{3}{7} = 20$$
$$150 \div 250 \times 20 = 12 + 8 = 20$$
$$225 \div 325 \times 20 = 13\tfrac{11}{13} + 6\tfrac{2}{13} = 20$$
$$300 \div 400 \times 20 = 15 + 5 = 20$$

It may be objected to this method that it does not make allowance for the angle of the belt; but in the above example the variation is but 1 in a distance of 6 feet between shafts—a difference which would scarcely be perceptible in the tension of the belt. If, however, the distance at which the shafts are to run is known, and it is desirable to be accurate, one of the cones may be set back of the lathe the proper distance, and the others may be turned by a tape line until the tension is equal upon all the pulleys. This works well in practice, and it is simple enough to be within the reach of any mechanic who is likely to have cones to make.

---

## FORMULÆ FOR CUTTING SCREW THREADS.

We are indebted to Mr. P. Golay, of Lima, Ohio, for the subjoined rule :—

FOR SINGLE GEARED LATHE.—Divide the number of threads you wish to cut (to the inch) by the pitch (number of threads to the inch) of the feed screw, and multiply the quotient by the number of teeth on the driving wheel, and the product is the number of teeth on the wheel driven.

*Examples.*—To cut 9 threads, pitch 5, driving wheel 25 teeth.

$\frac{9}{5}=1{\cdot}8\times25=45$ teeth on the wheel driven.

To cut $9\frac{1}{2}$ threads, pitch 5, driving wheel 20 teeth.

$\frac{9{\cdot}5}{5}=1{\cdot}9\times20=38$ teeth on the wheel driven.

To cut 10 threads, pitch 6, driving wheel 30 teeth.

$\frac{10}{6}=1{\cdot}666666\times30=50$ teeth on the wheel driven.

To cut $10\frac{1}{5}$ threads, pitch 5, driving wheel 25 teeth.

$\frac{10{\cdot}2}{5}=2{\cdot}04\times25=51$ teeth on the wheel driven.

FOR DOUBLE GEARED LATHE.—Divide the number of threads you wish to cut by the pitch of the feed screw, and multiply the quotient by the product of the number of teeth on the driving wheels; then any divisor that leaves no remainder to this product is the number of teeth on one of the wheels driven, and the quotient the number of teeth on the other wheel driven.

*Examples.*—To cut $9\frac{1}{2}$ threads pitch, 4, drivers 40 and 48 teeth.

$\frac{9\cdot5}{4}=2\cdot375\times40\times48=\frac{4560}{38}=120$. Or,

$\frac{9\cdot5}{4}=2\cdot375\times40\times48=\frac{4560}{60}=76$.

We get 38 and 120, or 60 and 76 for the number of teeth on the two wheels driven.

To cut $10\frac{1}{5}$ threads, pitch 4, drivers 24 and 30 teeth.

$\frac{10\cdot2}{4}=2\cdot55\times24\times30=\frac{1836}{36}=51$, we get 36 and 51 for the number of teeth on the two wheels driven.

To cut 3 threads, pitch 4, drivers 24 and 30 teeth.

$\frac{3}{4}=75\times24\times30=\frac{540}{20}=27$, we get 20 and 27 for the number of teeth on the two wheels driven.

To cut 10 threads, pitch 6, drivers 24 and 30 teeth.

$\frac{10}{6}=1\cdot66666\times24\times30=\frac{1400}{35}=40$ we get 35 and 40 for the number of teeth on the two wheels driven.

## CHAPTER XLI.

HOW TO LAY UP AN EIGHT-STRAND GASKET—TO TURN AN ELBOW—FLY-WHEELS FOR LONG SHAFTING—VELOCITY OF MECHANISM.

MANY an engineer who makes his own packing is contented to use a simple three strand loosely-plaited gasket, for all purposes whatever, from packing a simple governor valve stem, up to the piston rod, or air-pump bucket. Believing as we do that an eight strand gasket is much superior to the ordinary kind, that it will wear longer, is a better shape to conform to its situation, and that it requires less compression from the gland to bring it up against the rod, we have here illus-

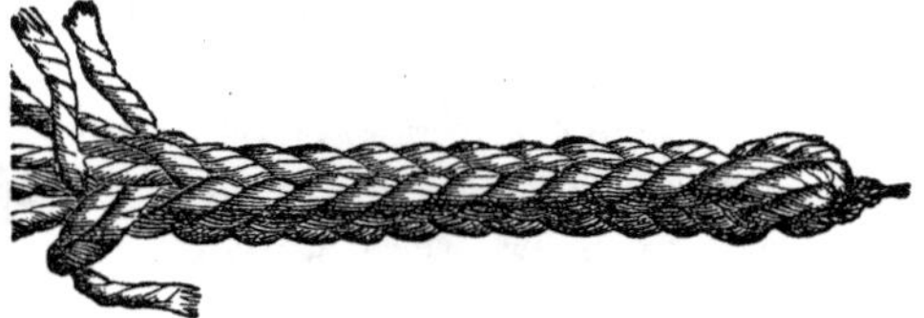

Fig. 79.

trated the gasket itself, and also the principle of laying it up. We have endeavored to make both the article and engravings simple and clear, and hope that the practical engineer will derive some benefit from our exposition.

Fig. 79 is the gasket as it appears finished, and

immediately below, in fig. 80, is given the first step toward forming it.

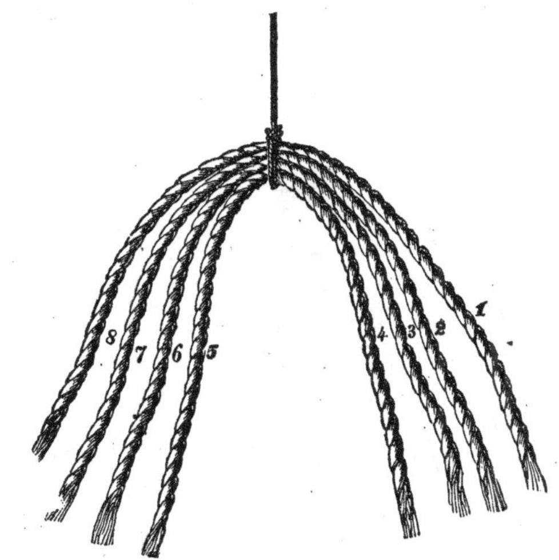


Fig. 80.

The operator takes eight strands, as shown. These are tied in the centre, and numbered, for the convenience of the reader, from 1 to 8.

In fig. 81 we have the two strands, 3 and 4, crossed under 5 and 6, and the thumb and forefinger of the left hand represented as closing upon these strands to retain them in place. In fig. 82 we have the real commencement of plaiting the gasket, and here is the point where the principle is first employed. This principle is that the strand, whatever one it may be the operator has hold of, *must pass under all the strands, and over two strands.* This is the key to the whole matter. It must also be borne in mind that *the top strand*

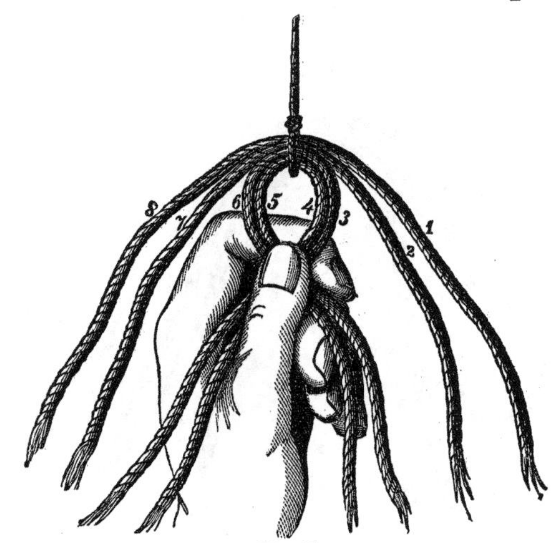


Fig. 81.

of all on each side, is the one to be taken hold of alternately. In fig. 82 the finger and thumb of the right hand are shown grasping the strand No. 8; the left hand being supposed to hold the crossed strands. Now look at the hand that grasps strand No. 8, it is inserted between strands 2 and 5, and is *behind* all the strands except 1 and 2, therefore when strand No. 8 is brought under all the strands except 1 and 2, and over strands 5 and 6, it will appear as in fig. 83, where strand No. 8 is shown drawn around, but not up to its place; the fingers of the right hand grasp it, and the left hand keeps the crossed ones, 3, 4, 5, 6, together.

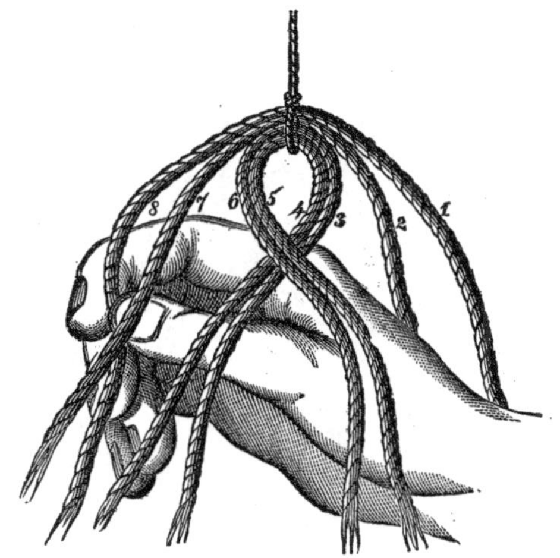


Fig. 82.

In the next figure, which is 84, the strand No. 8 is shown loosely drawn up to its place; the operator's hand going under all the strands for No. 1. This strand is to be brought *under* and behind all the other strands, and in between strands 3 and 4 where the hand enters, and thence over 3 and 8 as shown in fig. 85. Thus the principle of this gasket is illustrated, for it is only necessary to go between each alternate set of strands on either side—to take the topmost strand alternately, and to lay it over two strands, to make a hard, firm, and even piece of packing.

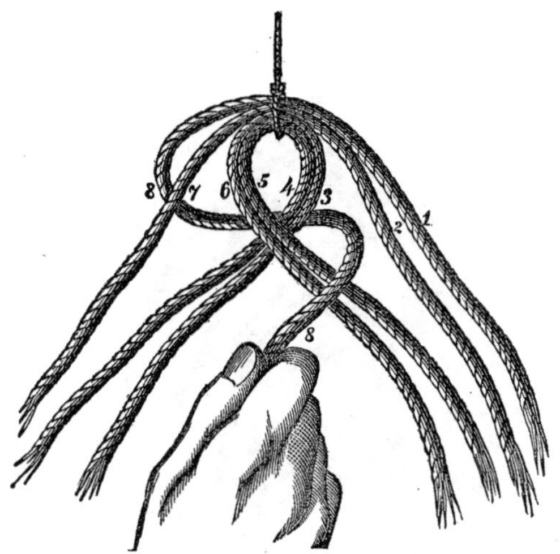


Fig. 83.

Some engineers prefer to use a central core of india-rubber, plaiting the strands over and about it, so that the rubber exerts its elastic force, but is not injured by the heat and grease of the machinery. This can be done very easily with the eight-strand gasket by merely allowing the rubber to occupy a central position between the strands, four on one side and four upon the other; the rubber must be cut square and to the proper size, and when it is overlaid with the strands it should be larger than the recess in the stuffing-box, so that

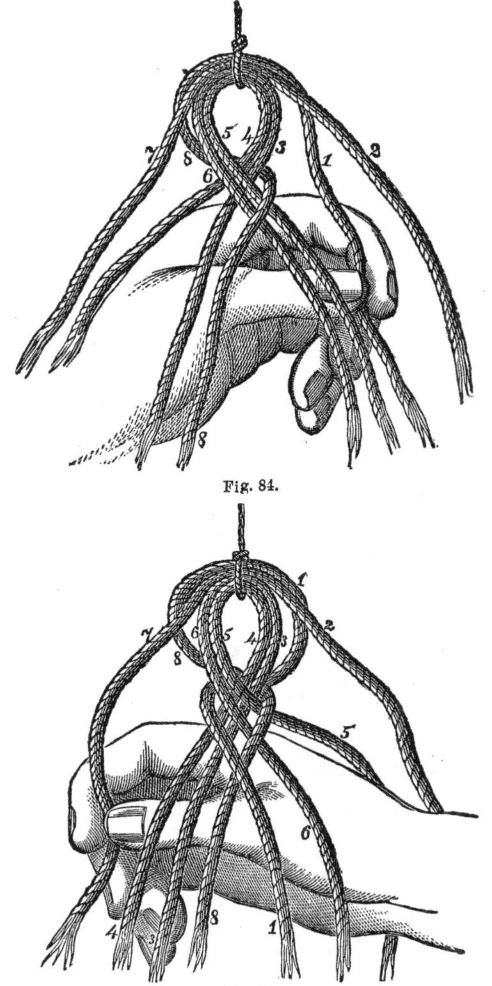

Fig. 84.

Fig. 85.

it will have to be compressed in order to get it in; it will then tend to cling about the rod and wear a long time with good usage.

Let the beginner not be discouraged at the first trial if he does not succeed, for the process is, in reality, a simple one, and inexperienced persons have made gaskets from these drawings at the first trial. The gasket should be laid up while reading this description, and we hope all points are made clear and simple, so that a little practice will, as in all other cases, render the braiding of a square gasket as easy as one of three strands.

---

## TO TURN AN ELBOW.

If we attempt to bend an elbow in a brass or copper pipe, unfilled with resin, it bulges out at

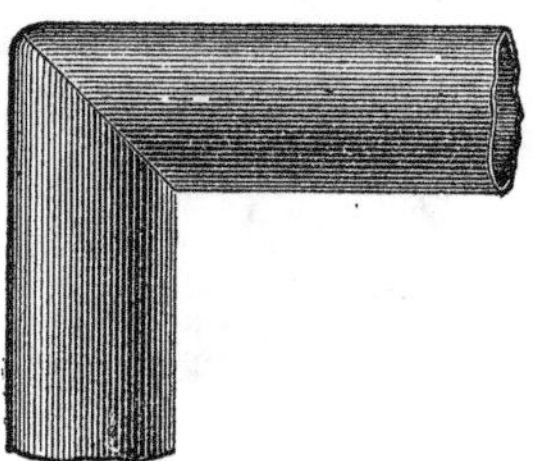

Fig. 86.

the sides and is dented in the middle, resulting in an unsightly piece of work. A very neat elbow can be made in a small pipe by cutting out a portion in the centre, bending the pipe over and soldering the parts, as shown in the previous engravings. If desired, a gusset can be put underneath the pipe, so as to make it stronger.

This is, of course, best adapted for light work, but it is at the same time quite strong. Where gas fitters' elbows cannot be readily obtained, it is an expeditious method of turning an elbow. For some purposes, also, pipes look better to have square corners. With this plan they are easily made so.

---

## FLY-WHEELS FOR LONG SHAFTING.

Long lines of shafting that communicate power to machines at a distance from the prime motor, spring and buckle greatly where the work is variable. The torsional or twisting strain, tending to wrench the shaft asunder, causes back-ash in the machinery driven, so that it runs fast and slow, or unevenly; this is often a source of great loss. The remedy is to put a moderately heavy fly-wheel on the extremity of the shaft, close to the hanger. This wheel takes up the strain and gives it out, or, in other words, equalizes the power, so that no change is perceptible. It is practiced in some of the Eastern cotton factories, and is found of great benefit.

---

## VELOCITY OF MECHANISM.

Fan blowers are frequently run with a velocity of 3,000 turns per minute, while the usual velocity of cotton spindles is between 6,000 and 7,000 turns per minute. These are the highest rotary velocities with which we are acquainted in ordinary mechanism, but M. Arago, in measuring the difference in the velocity of light while passing through air and through water, wished to give a revolving mirror a velocity of 8,000 rotations per second. This he was unable to do. With the most delicate and perfect arrangement of cog wheels he was able to impart only 1,000 revolutions per second to his mirror. M. Foucault, by substituting for cog wheels a delicate turbine acted on by a steam jet, raised the velocity to 1,500 turns per second. M. Arago, by removing the mirror and turning the spindle alone, achieved a velocity even by means of cog wheels, of 8,000 turns per second—equal to 480,000 turns per minute.

That spindle, therefore, turned 80 times, while an ordinary cotton spindle is turning once! This is the highest rotary velocity of which we have any account.

# CHAPTER XLII.

## VARIOUS USEFUL ITEMS.

RADIUS OF THE LINK MOTION.—The radius of the link in link motion for slide valves is struck from the centre of the shaft. The lead is supposed to remain the same, but it is generally increased slightly in cutting off shorter.

HOW CLOTH IS MADE WATERPROOF.—The method of making cloth waterproof consists in dissolving one ounce of alum and an equal quantity of the sugar of lead in one gallon of water and then allowing the sediment to fall to the bottom of the vessel. Now take the clear liquid, warm it, immerse the cloth in it, and hang it up to dry, after which it will repel water, but air will pass through it freely. It is not so perfectly water-proof as an india-rubber composition.

WEIGHT OF A CUBIC FOOT OF VARIOUS METALS.—A cubic foot of cast-iron weighs 450.55 pounds, of wrought iron 486.65, brass, 537.75 and fresh water 62.5.

A TIN VESSEL to hold one gallon beer measure should measure as follows:—Diameter of top 3¾, of the bottom 8¾ inches, height 7½ inches.

DOUBLE THREADED SCREWS.—A double threaded screw runs no faster through a nut than a single threaded one of the same pitch. Double or triple thread screws are the only means of cutting extraordinarily quick pitches on small rods or shafts.

GOLD POWDER.—A gold powder, according to "Cooley," is made by rubbing gold leaf with sulphate of potassa in crystals, the latter is afterward washed out. Another gold powder can be made by rubbing gold leaf on a marble slab with honey or molasses, and afterward washing out the molasses when the gold will sink to the bottom.

ARTIFICIAL GRINDSTONES.—An artificial grindstone can be made by the following formulæ, although the natural one is cheaper and better; washed silicious sand 3 parts, shellac 1 part, melt the lac and mould in the sand while warm. Emery may be substituted for sand. Used for razors and fine cutlery.

TO CLEAN BRASS CASTINGS.—A bath composed of one-part of hydrochloric acid to ten of water, will answer for cleaning brass castings. They should be dipped into an alkaline solution after being put into the acid, and washed and dried before being lacquered.

TO MAKE INDIAN INK.—Indian ink is a mixture of lamp-black and glue, with the addition of camphor and other substances in small quantities. It is said that the attempts to imitate it in this country and Europe have not been entirely successful. Many efforts have been made to obtain a suitable fluid vehicle for the carbon inks. Professor Traill says that an acetic solution of gluten answers the purpose. The gluten should be kept from 24 to 36 hours in water, and then be digested in acetic acid of specific gravity 1.033 to 1.034 in the proportion of 3 parts of gluten to 20 of acid. It is submitted to a gentle heat till a greyish white saponaceous fluid is obtained. Then 8 to 12 grains of the best lamp-black and 2 grains of indigo are incorporated with each fluid ounce of the liquid. Some cloves digested at first with the acid are thought to prevent mildew.

To Strike an Oval.—An "oval" is very easily made with a string and two stout pins. Tie the ends of a cord together, slip it over the pins loosely, and then place a sheet of paper under them. It is supposed that the pins have been driven into a table or board previously. Take a pencil and place it inside of the string, stretch the same out to one side and follow the pencil around the string on the paper.

Decomposition of Water.—When the vapor of water is passed through a gun barrel, maintained at a red heat in a furnace, the water is decomposed by the oxygen leaving the hydrogen and uniting with the iron of the barrel. The hydrogen which will then escape from the gun barrel is not explosive; it burns with a blue flame giving out a very intense heat, but emits very little light.

COLORIFIC VALUE OF HYDROGEN.—From a given weight of hydrogen gas under combustion, a greater quantity of steam can be generated than from an equal weight of any other known combustible. A pound of pure carbon will evaporate 12½ pounds of water and convert it into steam of 15 pounds pressure on the square inch. One pound of good Pennsylvania anthracite is capable of raising 9½ pounds of water at 212° Fah., into steam.

WEIGHT OF WATER.—The weight of one cubic foot of water is 62½ pounds. The weight of one gallon is about 8 pounds.

RED STAIN FOR WOOD.—To stain wood red take one pound of Brazil wood to one gallon of water, boil three hours with one ounce of pearlash, brush it hot on the wood and while hot brush the wood with a solution made with two ounces of alum in one quart of water.

SODA WATER.—Soda water is simply pure water saturated with carbonic acid, and is beneficial to most constitutions. The carbonic acid is usually obtained by pouring sulphuric acid upon marble dust. As the sulphuric acid is an active poison, effective measures must be taken to prevent any of it from getting into the beverage.

PROPORTION OF GRATE IN BOILERS.—The proportion of grate to heating surface is from one twenty-fifth to one thirty-fifth of the entire heating surface.

CHINESE WHITE COPPER.—Chinese white copper is made by taking copper 40.4, nickel 31.6, zinc 25.4, iron 2.6 parts.

TO GILD ON GLASS.—To gild letters on glass mix powdered gold with thick gum-arabic and powdered borax. Trace the device with this on the glass or china, and afterward bake it in a hot oven. The gum is thus burnt and the borax vitrifies and cements the gold to the ware. Powdered gold is made by rubbing gold-leaf with honey on a marble slab, then washing the composition when the honey is dissolved and the gold is precipitated in a powder.

BLACK JAPAN VARNISH IS THUS MADE: Pitch 50 lbs., dark gum amber 8 lbs., melt this and add linseed oil 12 gallons. Boil this and add 10 lbs., more gum amber previously melted and boiled with 2 gallons of linseed oil, 7 lbs. each of litharge and red-lead, and boil for two hours or until a little of the mass can be rolled into pills, then withdraw the fire and thin the varnish as required for use with turpentine.

TO CLEAN COINS.—Clean coins with dilute sulphuric acid; one part of acid in ten of water will answer very well. If there are dates upon them this will bring them out, if not, not. Nothing can bring out the date of coin which has been worn off. There are some old coin washers who have a simple method for bringing out dates—that is to manufacture them.

CUTTING OFF WITH THE LINK.—We cannot cut off at all points of the stroke with a link motion with economy. But at a certain point, which varies with the construction of the valve, the steam is not only cut off from the cylinder but shut up in it. In other words the exhaust is closed too soon and back pressure results.

RULE FOR HORSE-POWER.—Square the diameter of the cylinder and multiply the product by ·7854, this will give the number of inches area in the piston. Multiply the area by the pressure of steam and the number of feet the piston travels per minute. This must be divided by 33,000, which is supposed to be the standard for a horse-power. It seems that some misunderstand this simple matter, and ask whether a stroke is one movement of the piston or *two*. If the whole number of feet travelled by the piston in a minute be reckoned there can be no confusion. Of course, if a piston goes two feet in one movement through the cylinder, in coming back it travels two more, or four in one revolution. By the rule given, a 4 inch piston and 12 inches stroke, making 200 strokes in a minute, with 50 pounds pressure, is 3·111 horses-power. The square of the diameter is 4×4=16; which, multiplied by ·7854, gives 12·5664 as the piston area. This again multiplied by the steam pressure, 50 pounds, gives 628·3200, which, multiplied by the distance the piston travels per minute, 100 turns (or 200 feet), gives 125664·0000; this being divided by 33,000 pounds, a standard horse-power, gives 3·111 horses-power. We trust that is clear enough.

Green Bronze.—The green bronzes seen in the windows of stores are made so artificially. Bronzes steeped some days in a strong solution of common salt, if washed in water and allowed to dry slowly become permanently green. Or a strong solution of sugar with a little oxalic acid will produce the green color. A dilute solution of ammonia allowed to dry on the surface produces an evanescent green.

Weight of Cast-iron Balls and Rings.—The weight of any cast-iron ball is found by multiplying the cube of the diameter in inches by ·1377 the product will be the pounds avoirdupois. To find the weight of a cast-iron ring multiply the breadth of the ring added to the inner diameter by ·0074, and that again by the breadth and thickness. This will give the weight in cwts. of 112 lbs., nearly.

To find the Specific Gravity of Metals.—To ascertain the specific gravity of a piece of metal, suspend the piece below one scale of a balance by a fine filament of raw silk, and weigh it. Then lower the balance so as to allow the metal to be submerged in water and weigh again. Divide the weight in air by the difference between the two weights, and the quotient is the specific gravity. The water must be pure.

Centrifugal Force.—To find the centrifugal force of any body, multiply the square of the number of revolutions per minute by the diameter of the circle in feet, and divide the product by 5·780. The quotient is the centrifugal force in terms of the weight of the body.

Inventor of Safety-valve.—The safety-valve was first invented by Papin in 1700, who applied it to a cooking machine for digesting bones, meats, etc.

To Bronze Cast-iron.—To bronze cast-iron, clean it thoroughly and dip it into a solution of sulphate of copper. The sulphate of copper is dissolved in water.

To bend Files.—Files are not manufactured bent to order, but could be if desired. Files can be bent by heating them to a dull red and striking them with a wooden mallet on a block of wood. Re-heat the file as high as possible without burning it, and plunge into cold water.

TO TAKE STAINS FROM CUTLERY.—Take the softest side of a razor strap, put rouge powder or crocus on it, and rub the cutlery rapidly with it. This will remove ordinary stains, but if they are rusted in, it is better to send the goods to a cutler to be refinished.

ALLOY FOR JOURNAL BOXES.—A good alloy for strong brass (or composition) boxes to carry heavy shafts is made in the following proportions: Tin, 2½ ounces; zinc, ½ of one ounce to 1 pound of copper. Common yellow brass is made harder by the addition of ¼ of an ounce to the pound; lead, in the same proportion, makes it more ductile, so that it casts sharper in the mould.

HOW COTTON WASTE TAKES FIRE SPONTANEOUSLY.—When cotton waste or shavings are saturated with oil, a large surface is exposed to the action of the air, and if the oil has the property of absorbing oxygen it may absorb the gas so rapidly as to take fire. This is the way in which spontaneous combustion takes place. As petroleum naptha does not absorb oxygen, it never takes fire by spontaneous combustion.

WHAT CAUSES A STEAM-WHISTLE TO SOUND.—Steam causes the whistle to vibrate rapidly, that is what gives the sound. The steam strikes the thin edge percussively, or like a hammer, and that is what makes the vibrations. The pitch or note of all musical sounds is determined by the number of vibrations occurring in a given time.

PROPORTION OF WATER FORCING PISTONS.—In forcing water from a small·cylinder into a large one, the distances through which the two pistons move are in inverse proportion to their areas. The easiest way to find the area of a circle is to multiply the square of the diameter by ·7854. The area of a 3 inch piston is 7, and of a 12 inch 113. Therefore, in forcing a small piston down 12 inches, it will raise the larger one 113 : 7 :: 12 : .74, say $\frac{3}{4}$ths of an inch.

THE MOMENTUM OF A MOVING BODY is its mass multiplied into its velocity, while the *vis viva* is is one-half the mass multiplied into the square of the velocity. Momentum is a mere term employed in certain mathematical processes with no corresponding quantity in nature, but *vis viva* or "live strength," is the actual force exerted by any moving body—the sum of the resistance required to bring the body to a state of rest.

TO MAKE HYDROGEN GAS.—To procure hydrogen gas, dilute 3 pounds of oil of vitriol with 24 pounds of water, and dissolve in it 2 pounds of zinc. All the apparatus required is an air-tight glass or lead vessel, with a pipe inserted air-tight in the top to carry off the gas. Navigating balloons will always be impracticable, for the reason that a balloon which will float an engine in the air must be too bulky to be moved with any but the most moderate velocity through the air. Fire shells have long been made far more efficient than Greek fire, or any other liquid.

TO FIND THE WEIGHT OF CASTINGS.—The weight of castings can be found by multiplying the width in quarter inches by the thickness in one-eighth inches. The result is in pounds per foot of length. We have never tested this rule but it is said to be a good one.

WELDING POINT OF IRON.—The welding point or iron is from 12,000 to 13,100 degrees. Cast-iron melts at from 17,000 to 20,000 degrees of heat.

TO BROWN A GUNBARREL.—A tincture of iodine diluted with half its bulk of water will produce a superior brown tint on the barrel of a fowling-piece.

HARD ALLOY OF COPPER AND TIN.—One thousand copper to fourteen tin is said to be the alloy from which the ancients made tools which would cut hard metals.

TO STICK BRASS LETTERS ON GLASS.—Resin 150 parts, wax 30, burnt ochre 30, and calcined plaster 2 parts. Apply warm.

BURNT TALLOW.—Burnt tallow or grease cannot be restored to its original condition. When subjected to high temperatures animal and vegetable fats and oils are completely changed in their characteristics. At a red heat they are converted into inflammable gases.

TO MAKE BRICK WATERPROOF.—To exclude dampness from brick-work, varnish it with a coating made in the the proportion of mixing 8 lbs. of linseed oil with 1 lb. of sulphur, and heating to 278°. We know of no better paint than that made of red lead and linseed oil.

LOCOMOTIVE CYLINDERS.—Locomotive cylinders are made of various diameters, the most general are from 18 to 20 inches by the same length of stroke. Some very heavy locomotives have 22 inch cylinders, and we think there is one in this country which has 26 inch cylinders.

# INDEX.

PAGE
Abuse of files... 126
Abuses of chucks... 62
Adhesion of a belt... 245
Adhesion of belts, use of sawdust or resin for increasing... 240
Adjustable tool for holes of varying diameters... 25, 26
Air-pumps... 184
Alloy for journal boxes... 267
Alloy of copper and tin, hard... 269
Aluminum bronze... 140
Anthracite coal as fuel... 221
Arago's experiments in velocity of mechanism... 258
Ashes around the grate... 200
Auger pod... 52

Balanced slide valves... 167
Balls and rings, weight of cast-iron... 265
Bar and cutter... 117
Beading tool... 101
Bead on a side pipe, turning or moulding... 101
Beam engines, fastest... 151
Beam engines, piston and speed of... 147
Bearing surfaces... 185
Belts, adhesion of... 245
Belt passed over a pulley, adhesion and tension of... 245
Belt, power and speed of... 245
Belt, pressure of... 245
Belt, speed and power of... 245
Belt, strength of... 245
Belts... 247
Belts at Cincinnati bridge, power and performance of... 243
Belts, ends of... 241
Belts, holes in... 241
Belts, lacing... 240
Belts, John A. Roebling on the power of... 242
Belts, relative resistance of... 239
Belting... 240
Belting, horse-power of... 237
Bits... 51

PAGE
Bit, rose... 29
Black Japan varnish... 263
Blacksmith... 109
Blacksmiths, use of callipers by them... 114
Boilers, explosion of... 200
Boilers, faulty construction of... 211
Boilers, frequency of explosions... 202
Boilers, field for improvement in... 221
Boilers, grate in... 263
Boilers, imperfect setting of... 216
Boiler, is yours safe?... 205
Boilers, leaks in... 209
Boilers, rubbish in... 194
Boilers, starting fires under... 217
Boilers, theories of explosion of... 202
Boilers, trouble incident to... 214
Bonnets, steam chest with... 158
Boxes for fan blowers... 140
Boring bar and cutter... 58, 60
Boring tool... 26
Boring tools... 46
Boring steam cylinders and hollow ware... 64
Bourne's catechism quoted... 154
Brass... 139
Brass castings, to clean... 260
Brass letters, to stick on glass... 269
Brass, speed in cutting... 41
Brass, tool for boring... 49
Brass, tools for turning... 101
Breaking drills... 35
Brick, to make waterproof... 270
Bridges, girders... 128
Brown a gunbarrel, to... 269
Bronze, green... 264
Bronze cast-iron... 266
Burning iron castings... 130
Burnt tallow... 270

"Cahawba" steamer... 208
Callipers, how to use... 113
Carbon, combination with oxygen 221
Carbonic acid, absorption of by steam... 224
Carlyle's experiments in condensation... 175

PAGE
Case-hardening.......... 140
Cast gears.......... 237
Cast-iron, to bronze.......... 266
Castings, to find the weight of.......... 269
Centre made with centre punch... 74
Centre or tit drill.......... 30
Centre punch.......... 74
Centres enlarged with a countersink.......... 74
Centres, knocking off the.......... 42
Centres, spare the.......... 108
Centres should be drilled.......... 73
Centrifugal force.......... 266
Chasers on taps, use of.......... 124
Chinese white copper.......... 263
Chisels, angle for cutting with.......... 18
Chucks.......... 44
Chucks, abuses of.......... 61
Cincinnati bridge, leather belts at 243
"City of Buffalo," engine of.......... 148
Clayey substances, expansion of... 228
Clean coins, to.......... 264
Clean brass castings, to.......... 260
Cloth, how to make waterproof.... 259
Chucking work in lathes.......... 44
Coins, to clean.......... 264
Cold weather and steam-engines... 196
Collars in a shaft.......... 131
Colorific value of hydrogen.......... 262
Common straight cutting off tool.. 96
Composite drills.......... 31
Compression, theory of.......... 165
Condensation of steam in long pipes.......... 175
Cone pulleys for given velocities... 246
Cone, velocity of.......... 246
Connection of slide valves.......... 170
Conservatism among mechanics... 69
Constituents of heat.......... 227
Countersink.......... 74
Cored holes in castings, making smooth.......... 115
Cloth waste, how it takes fire spontaneously.......... 267
Copper and tin, hard alloy of.......... 269
Counter-borer drills, variety of..... 24
Counter-borers.......... 55, 56, 57
Counter-borer.......... 22
Counter-boring in lathes.......... 32
Countersinks for enlarging centres 75
Crank shaft of "Golden Gate"...... 94
Crank shafts, heavy.......... 94
Cranks for marine engines.......... 57
Crystals of cast metals, cutting.... 47
Currents of steam, action of.......... 226
Curve, working out.......... 92
Cutlery, to take stains off.......... 267
Cutter.......... 76
Cutter, a useful.......... 28
Cutter, feed and speed of.......... 61
Cutter for boring cylinders.......... 61
Cutter of drill.......... 18
Cutting off tool.......... 96

PAGE
Cutting off the link.......... 264
Cutting tools, speed of.......... 37
"C. Vanderbilt" performance of engines.......... 149
Cylinder head, turning a moulding on.......... 101
Cylinders, boring.......... 61
Cylinders, locomotive.......... 270

Dead centre of lathe.......... 94
Decomposition of water.......... 261
Defect in steam-engines.......... 161
Deep holes, drilling.......... 18
Defective iron castings.......... 127
Depth of teeth.......... 236
Derangement of steam-engines..... 194
Diamond point tool.......... 75, 79
Difficulties with engines.......... 163
"Doctor," use of in turning heavy shafts.......... 82
Dog.......... 74
Double beat valves.......... 151
Double threaded screws.......... 260
Drill.......... 13, 14, 16
Drill and its office.......... 13
Drill, badly made.......... 16
Drill, cutters of.......... 18
Drill for boring tube sheets.......... 28
Drill for counter-boring in lathes.. 32
Drill for drilling tube sheets of surface condensers.......... 31
Drill, grinding.......... 18
Drill, lipped.......... 17
Drill, Morse.......... 21
Drill, pin.......... 22
Drill, plain.......... 17
Drill, proper mode of making.......... 19
Drill shank.......... 33
Drill shank, true shape of.......... 34
Drill, turned.......... 31
Drills, complexity in form of.......... 35
Drills, composite.......... 31
Drills, counter-borers.......... 24
Drills, great variety of.......... 33
Drills, Manhattan Fire-arms Co.'s 22
Drills, twist.......... 19
Drilling and turning glass.......... 136
Drilling deep holes.......... 18
Drip pans, use of.......... 136
Driving cone, diameter of.......... 247
Driving cone, velocity of.......... 246
Dust and cinders, accumulation of 209
Dynamometer.......... 238, 239

Eccentric.......... 156
Eccentric, position of.......... 156
Economical working of machinery 185
Economy in fuel.......... 180
Economy, want of, in working steam-engines.......... 161
Edge of a filed tool.......... 101
Eight-strand gasket.......... 250
Elbow, to turn.......... 256

PAGE
Electricity and steam-boilers........ 218
Emery, use of............................ 79
Ends of belts.............................. 241
Engines, difficulties with............. 163
Engine of "City of Buffalo"......... 148
Engines of "C. Vanderbilt"........... 148
Engines of the "Jesse Hoyt"....... 151
Engines of "Metropolis".............. 149
Engines of "Mississippi"............ 149
Engines of "New World"............ 150
Engines of the "Stockton"........... 151
Epicycloidal curve, a true............ 234
Epicycloidal teeth................ 233, 234
Exhausting "one sided"............... 163
Exhaust valve............................ 152
Expansion, laws of..................... 227
Expansion of valve..................... 159
Expansion, ratio in different metals.......................................... 229
Expansion when uncompensated for in machinery...................... 229
Experiments with tools............... 64
Explosions of steam-boilers... 200, 201

Fats, sulphuric acid in................ 192
Faulty construction of steam-boilers.................................... 211
Favre and Silberman's experiments ....................................... 221
Feed and speed of cutter.............. 61
Feed pumps................................ 162
Feed pumps, strain on when frozen 197
Fibres of wrought-iron cutting..... 47
Field for improvement in steam-boilers.................................... 221
Filed, tools which cannot be......... 100
Files, abuse of............................ 126
Files, to bend............................. 266
Finishing or polishing a shaft...... 79
Finishing tool............................. 80
Finishing tools............................ 46
Finishing tool, use of.................. 80
Fire surface............................... 162
Fires, starting under boilers......... 217
Flue sheets, holes in................... 24
Fly wheels for long shafting......... 257
Foucault's experiments in velocity of mechanism.......................... 258
Forgings, good........................... 111
Form of drill shank..................... 33
Formula for cutting screw threads 248
Friction of slide valves................ 169

Gallon measure, size of............... 259
Gasket, to lay out an eight-strand 250
Gears ........................................ 231
Gears, cast................................. 237
Gearing absorbs less power than belts......................................... 239
German brass............................. 139
Gib-heads to keys....................... 121
Gibs.......................................... 188

PAGE
Gild on glass, to......................... 263
Girders of bridges....................... 128
Glass, drilling and turning........... 136
Glass, to gild on......................... 263
Glass, to stick brass letters on...... 269
"Golden City," performance of engine.................................... 147
"Golden Gate," crank shaft of...... 94
Gold powder.............................. 260
Grate in boilers.......................... 263
Green bronze............................. 264
Grinding tools............................ 98
Grinding a drill.......................... 18
Grindstones, artificial.................. 260
Gunbarrel, to brown.................... 269

Half set of drills......................... 21
Handy tool................................ 115
Hard alloy for copper and tin....... 269
Harsh belts, to soften.................. 242
Heat, best mode of transferring.... 224
Heat, constituents of................... 227
Heat, loss of in steam-boilers........ 221
Heat, radiation of....................... 222
Heating surface.......................... 162
Heat, transfer of......................... 223
Heavy crank shafts..................... 94
Heavy shafts, use of a "doctor" in turning.................................... 82
Heavy work............................... 57
Holes, boring of in metal............. 14
Holes, boring with bar and cutter 117
Holes, economy in boring............ 47
Holes in belts............................ 241
Holes in castings, making smooth 115
Holes in metals, badly bored........ 17
Holes, tapping............................ 125
Hole 20 inches in diameter, to cut 58
Hollow work for boring................ 64
Horse-power, a.......................... 237
Horse-power of belting................ 237
Horse-power, rule for.................. 264
How to set a slide valve.............. 153
Hydrogen, colorific value of......... 262
Hydrogen gas, to make................ 268

Idling over work......................... 61
Improvement in steam-boilers...... 221
Improperly set valve................... 157
Impurities in iron........................ 111
Indicator.................................... 159
Indicating lead........................... 159
Indicators, location of................. 225
Indian ink.................................. 260
Inside and out, turning a job....... 93
Inventor of safety valve.............. 266
Iron and brass mixture................ 139
Iron castings, burning................. 130
Iron castings, defective................ 127
Iron forgings.............................. 109
Iron, over-heated........................ 111
Iron, speed run in cutting............ 38

PAGE
Iron, welding point of .... 269
Irregularly shaped work.... 46
Is your boiler safe?.... 205

Japan varnish.... 263
"Jesse Hoyt," engines of.... 151
Jobber's set of drills.... 21
"John Faron," steamer.... 224
Journal boxes, alloy for.... 267
Journal oiling.... 135
Journals of steam-engines, convex 187

Keying wheels on shafts.... 119
Key, the hold of.... 121
Key, the principle of.... 119

Lacing belts.... 240
Lathe, double geared.... 249
Lathe, management of.... 71
Lathe, single geared.... 248
Lathe work.... 37
Lathe, chucking work in.... 44
Lathes, the points to be considered in running.... 38
Laws of expansion.... 227
Lead, expansion of.... 228
Lead pipes used for hot water.... 228
Lead indicator.... 159
Lead of slide valve.... 158
Leaks in boilers.... 209
Learn to forge your own tools.... 104
Leather bands.... 237
Length of the rod, to find.... 155
Lightning rod, turning the forked end.... 88
Link, cutting off.... 264
Link motion.... 158
Link motion, radius of.... 259
Lipped drill.... 17, 18
Location of steam gages and indicators.... 225
Locomotive, crank shaft for.... 94
Locomotive cylinders.... 270
Long shafts, use of spring tool in turning.... 87
Loss of power of steam in practice 161
Lubricating the steam-engine.... 189

Machinery, when expansion is uncompensated for, in.... 229
Machinist's drill.... 14, 15
Machinist's set of drills.... 21
Manchester association for inspection of boilers.... 200
Man-hole plate.... 207
Manhattan Fire-arms Co.'s drill.... 22
Manipulation of metals.... 138
Manual dexterity.... 106
Marine engines, large crank shafts for.... 94
Mariotte's laws of expansion.. 157, 159
Mechanical accuracy, cost of.... 164
Mechanics, conservatism of.... 69

PAGE
Mechanism, velocity of.... 258
Metals, expansion of.... 227
Metals, manipulation of.... 138
Metals, ratio of expansion in.... 229
Metals, to find the specific gravity of.... 266
Metals, weight of a cubic foot of... 259
Metal-working, perfection of.... 13
"Metropolis," engines of.... 149
Miscellaneous tools and processes. 104
"Mississippi," engines of.... 149
Momentum of a moving body.... 268
Morse drill.... 21
Moulding, turning on a side pipe.. 101
Moving body, momentum of.... 268

New engines should be inspected.. 196
"New World," engines of.... 150

Oil cups.... 135
Oil entering a steam cylinder against pressure.... 198
Oil, waste of.... 189
Oval, to strike an.... 261
Overheated iron.... 111

Packing steam pistons.... 180
Parallel holes.... 54
Penn's invention for slide valve.... 169
Pin drill.... 22
Piston, removing from the cylinder 180
Piston, speed of beam engines.... 147
Pistons, proportion of water forcing.... 268
Pistons without packing.... 183
Pitch lines, to lay out teeth on.... 234
Planers, time run by.... 40
Plain drill.... 17
Planing machines, V-shaped slides 186
Planing tool.... 19
Pod augur.... 52
Point of tool, position of in finishing.... 81
Poppet valve.... 152
Portable engines.... 162
Position of tool, importance of in cutting.... 81
Power absorbed in operating the slide valve.... 167
Power of a belt.... 245
Pressure of a belt.... 245
Pressure on a slide valve.... 173
Propeller shafts heated.... 140
Proportion of grate in boilers.... 263
Proportion of water forcing pistons 268
Pulley, belt or rope passing over... 245

Radius of the link motion.... 259
Railroad tracks, allowance for expansion.... 230
Ratio of expansion in different metals.... 229
Red stain for wood.... 262

PAGE
Resistance, relative, of belts........ 239
Resinous deposit in tubes of boilers 215
Rimmers........................ 60, 116
Rings, weight of cast-iron............ 263
Roebling, John A., on the power of belts.............................. 242
Rose bit.................................. 29
Rod, to find the length of........... 155
Rough forgings.......................... 109
Roughing tools...................... 46, 75
Round nose filleting tool............ 98
Round nose tool......................... 77
Rule for horse-power................. 264

Safety-valve, inventor of............ 266
Salt water in boilers.................. 208
Scale on the boiler flues.... ........ 207
Scale, removing the.................... 111
Science of steam engineering........ 142
Scraped surfaces........................ 132
Scraper.................................... 133
Screw engines............................ 187
Screw engines, crank shafts for.... 94
Screws, double threaded.............. 260
Screw threads, cutting................ 248
Screw threads, formula for cutting 248
Scroll chucks............................. 45
Seely, Prof................................ 224
Setting valve............................. 158
Setting of boilers........................ 216
Shaft, finishing or polishing......... 79
Shaft, roughing off a large........... 77
Shafts...................................... 112
Shafts, keeping wheels on............ 119
Shafts, use of a "doctor" in turning........................................ 82
Shafting, fly wheels for long......... 257
Shafting for pulleys, turning........ 85
Shafting turning in the lathe with concentric cutters.................... 85
Shanks of boring tool.................. 49
Shell arbor....................... 53, 54, 55
Shrink collars in a shaft.............. 131
Side cutting tool......................... 82
Side tool................................... 79
Silver solder..................... ........ 138
Sleeve, cast-iron, with set screws.. 86
Slide valve................................ 164
Slide valve, friction of................ 169
Slide valve, pressure on............... 173
Slide valve, to set....................... 153
Slide valve, virtue in the mechanical adjustment........................ 170
Slide valve, working.................... 171
Slide valves............................... 162
Slide valves, balanced.................. 167
Slide valves, connection of........ .. 170
Slide valves, designing the outward form of.......................... 166
Sliding of teeth on each other...... 235
Soda water................................ 262
Solid crank shaft........................ 95
Solids, amount of expansion of...... 229

PAGE
Spare the centres........................ 108
Specific gravity of metals, to find.. 266
Speed and feed of cutter............. 61
Speed of cutting tools.................. 37
Speed of a belt........................... 245
Speed with which tools cut iron and brass.............................. 13
Spelter for uniting iron and brass 139
Spindle, squared........................ 34
Spontaneous combustion............. 267
Spring tool................................ 87
Squared centre.......................... 74
Square-nosed tool................... 90, 96
Square rimmers.......................... 117
Stain, red, for wood.................... 262
Stains, to take from cutlery......... 267
Starting fires under boilers.......... 217
Stationary engines...................... 162
Stay and socket bolts of water-spaces.................................... 209
Steam and the steam-engine......... 142
Steam-boilers and electricity........ 218
Steam-boilers, loss of heat in.. ..... 221
Steam chest with bonnets............ 158
Steam, computing the actual volume of............................... 161
Steam cylinder, boring................ 64
Steam-engine............................. 142
Steam-engine, lubricating............ 189
Steam-engines, defective.............. 161
Steam-engines in cold weather...... 196
Steam-engineering, science of....... 142
Steam gages and indicators, location of.................................... 225
Steam generated from water........ 206
Steam in long pipes, condensation of.......................................... 175
Steam pistons, packing................ 180
Steam, pressure of...................... 157
Steam-whistle, what causes it to sound...................................... 267
Steel, annealing of...................... 13
"Stockton," engines of................ 151
Strike an oval, to........................ 261
Stubb's rimmer........................... 117
Sulphuric acids in fats................. 192

Tallow, burnt............................. 270
Tap, finishing in the lathe............ 124
Tap, the office of........................ 122
Taps and their construction......... 122
Taps, form of............................. 123
Tapping holes............................. 125
Teeth of wheels badly constructed 232
Teeth of wheels, importance of, construction of........................ 232
Teeth, to find the shape of........... 234
Teeth, width and depth of............ 236
Tin and copper, hard alloy of....... 269
Tit or centre drill........................ 30
Tool by which holes in flue sheets can be made of any size........... 24
Tool to cut large holes..... .......... 58

PAGE
Tools, experiments with... 64
Tools, learn to forge your own... 104
Tools, speed of to cut iron and brass... 13
Tools, turning... 69
Tools which cannot be filed... 100
Tooth, proper form of... 233
Tractive force of leather bands... 237
Transferring heat, best mode of... 224
Transmission of power, mode of estimating... 238
Trip hammers... 110
Troubles incident to steam-boilers 214
Tube sheet borer... 26
Tube sheets, drill for boring... 28
Turn an elbow, to... 256
Turned drill... 31
Turning a job inside and out at the same time... 93
Turning shafting... 85
Turning tools... 69
Twist drills... 19, 20

Useful items, various... 259

Valve, improperly set... 157
Valve, poppet... 152
Valve-seats, making... 55
Valve setting... 158
Valve-stem, brazing with silver solder... 138
Valves and piston, lubricating... 190
Valve, double beat... 151
Valve setting... 159
Valves, taking apart... 152
Varnish, black Japan... 263
Velocity of mechanism... 258
Vitreous scale on metal wheels... 231

PAGE
Water, decomposition of... 261
Water forcing pistons, proportion of... 268
Water passages of steam-engines.. 162
Water-proof, to make brick... 270
Water-proof, to make cloth... 259
Water, proper state of, in boiler... 206
Water, too little in boiler... 206
Water, too much in boiler... 206
Water, use of, with lathe... 103
Water, weight of... 262
Weight of a cubic foot of various metals... 259
Weight of castings, to find... 269
Weight of cast-iron balls and rings 265
Weight of water... 262
Welding point of iron... 269
Wheels, rattling... 232
Wheels, vitreous scale on... 231
White copper, Chinese... 263
Width and depth of teeth... 236
Winter, steam-engines in... 196
Wire gauge set of drills... 21
Wire rope passed over a pulley, adhesion and tension of... 245
Wood, red stain for... 262
Woody substances, expansion of... 228
Working a slide valve... 171
Wrought-iron, hard tools for turning... 102
Wrought-iron, speed run in cutting... 40
Wrought-iron, tool for boring... 40
Wrought-iron, use of a "doctor" in turning... 84

Yellow brass... 139

Zinc in mixing... 139

THE END.

# CATALOGUE

OF

# PRACTICAL AND SCIENTIFIC BOOKS,

PUBLISHED BY

## HENRY CAREY BAIRD,

INDUSTRIAL PUBLISHER,

No. 406 WALNUT STREET,

PHILADELPHIA.

☞ Any of the Books comprised in this Catalogue will be sent by mail, free of postage, at the publication price.

☞ This Catalogue will be sent, free of postage, to any one who will furnish the publisher with his address.

**ARMENGAUD, AMOUROUX, AND JOHNSON.—THE PRACTICAL DRAUGHTSMAN'S BOOK OF INDUSTRIAL DESIGN, AND MACHINIST'S AND ENGINEER'S DRAWING COMPANION:** Forming a complete course of Mechanical Engineering and Architectural Drawing. From the French of M. Armengaud the elder, Prof. of Design in the Conservatoire of Arts and Industry, Paris, and MM. Armengaud the younger and Amouroux, Civil Engineers. Rewritten and arranged, with additional matter and plates, selections from and examples of the most useful and generally employed mechanism of the day. By WILLIAM JOHNSON, Assoc. Inst. C. E., Editor of "The Practical Mechanic's Journal." Illustrated by 50 folio steel plates and 50 wood-cuts. A new edition, 4to. . $10 00

**ARROWSMITH.—PAPER-HANGER'S COMPANION:** A Treatise in which the Practical Operations of the Trade are Systematically laid down: with Copious Directions Preparatory to Papering; Preventives against the Effect of Damp on Walls; the Various Cements and Pastes adapted to the Several Purposes of the Trade; Observations and Directions for the Panelling and Ornamenting of Rooms, &c. By JAMES ARROWSMITH, Author of "Analysis of Drapery," &c. 12mo., cloth . . . . . . . . . . $1 25

**BAIRD.—THE AMERICAN COTTON SPINNER, AND MANAGER'S AND CARDER'S GUIDE:**

A Practical Treatise on Cotton Spinning; giving the Dimensions and Speed of Machinery, Draught and Twist Calculations, etc.; with notices of recent Improvements: together with Rules and Examples for making changes in the sizes and numbers of Roving and Yarn. Compiled from the papers of the late ROBERT H. BAIRD. 12mo. . . . $1 50

**BAKER.—LONG-SPAN RAILWAY BRIDGES:**

Comprising Investigations of the Comparative Theoretical and Practical Advantages of the various Adopted or Proposed Type Systems of Construction; with numerous Formulæ and Tables. By B. Baker. 12mo. . . . . . . $2 00

**BAKEWELL.—A MANUAL OF ELECTRICITY—PRACTICAL AND THEORETICAL:**

By F. C. BAKEWELL, Inventor of the Copying Telegraph. Second Edition. Revised and enlarged. Illustrated by numerous engravings. 12mo. Cloth . . . . . $2 00

**BEANS.—A TREATISE ON RAILROAD CURVES AND THE LOCATION OF RAILROADS:**

By E. W. BEANS, C. E. 12mo. (In press.)

**BLENKARN.—PRACTICAL SPECIFICATIONS OF WORKS EXECUTED IN ARCHITECTURE, CIVIL AND MECHANICAL ENGINEERING, AND IN ROAD MAKING AND SEWERING:**

To which are added a series of practically useful Agreements and Reports. By JOHN BLENKARN. Illustrated by fifteen large folding plates. 8vo. . . . . . . . $9 00

**BLINN.—A PRACTICAL WORKSHOP COMPANION FOR TIN, SHEET-IRON, AND COPPER-PLATE WORKERS:**

Containing Rules for Describing various kinds of Patterns used by Tin, Sheet-iron, and Copper-plate Workers; Practical Geometry; Mensuration of Surfaces and Solids; Tables of the Weight of Metals, Lead Pipe, etc.; Tables of Areas and Circumferences of Circles; Japans, Varnishes, Lackers, Cements, Compositions, etc. etc. By LEROY J. BLINN, Master Mechanic. With over One Hundred Illustrations. 12mo. $2 50

**BOOTH.—MARBLE WORKER'S MANUAL:**
Containing Practical Information respecting Marbles in general, their Cutting, Working, and Polishing; Veneering of Marble; Mosaics; Composition and Use of Artificial Marble, Stuccos, Cements, Receipts, Secrets, etc. etc. Translated from the French by M. L. Booth. With an Appendix concerning American Marbles. 12mo., cloth . . $1 50

**BOOTH AND MORFIT.—THE ENCYCLOPEDIA OF CHEMISTRY, PRACTICAL AND THEORETICAL:**
Embracing its application to the Arts, Metallurgy, Mineralogy, Geology, Medicine, and Pharmacy. By James C. Booth, Melter and Refiner in the United States Mint, Professor of Applied Chemistry in the Franklin Institute, etc., assisted by Campbell Morfit, author of "Chemical Manipulations," etc. Seventh edition. Complete in one volume, royal 8vo., 978 pages, with numerous wood-cuts and other illustrations. $5 00

**BOWDITCH.—ANALYSIS, TECHNICAL VALUATION, PURIFICATION, AND USE OF COAL GAS:**
By Rev. W. R. Bowditch. Illustrated with wood engravings. 8vo. . . . . . . . . . $6 50

**BOX.—PRACTICAL HYDRAULICS:**
A Series of Rules and Tables for the use of Engineers, etc. By Thomas Box. 12mo. . . . . . $2 00

**BUCKMASTER.—THE ELEMENTS OF MECHANICAL PHYSICS:**
By J. C. Buckmaster, late Student in the Government School of Mines; Certified Teacher of Science by the Department of Science and Art; Examiner in Chemistry and Physics in the Royal College of Preceptors; and late Lecturer in Chemistry and Physics of the Royal Polytechnic Institute. Illustrated with numerous engravings. In one vol. 12mo. . $2 00

**BULLOCK.—THE AMERICAN COTTAGE BUILDER:**
A Series of Designs, Plans, and Specifications, from $200 to to $20,000 for Homes for the People; together with Warming, Ventilation, Drainage, Painting, and Landscape Gardening. By John Bullock, Architect, Civil Engineer, Mechanician, and Editor of "The Rudiments of Architecture and Building," etc. Illustrated by 75 engravings. In one vol. 8vo. . . . . . . . . . . $3 50

**BULLOCK.—THE RUDIMENTS OF ARCHITECTURE AND BUILDING:**

For the use of Architects, Builders, Draughtsmen, Machinists, Engineers, and Mechanics. Edited by JOHN BULLOCK, author of "The American Cottage Builder." Illustrated by 250 engravings. In one volume 8vo. . . . $3 50

**BURGH.—PRACTICAL ILLUSTRATIONS OF LAND AND MARINE ENGINES:**

Showing in detail the Modern Improvements of High and Low Pressure, Surface Condensation, and Super-heating, together with Land and Marine Boilers. By N. P. BURGH, Engineer. Illustrated by twenty plates, double elephant folio, with text. $21 00

**BURGH.—PRACTICAL RULES FOR THE PROPORTIONS OF MODERN ENGINES AND BOILERS FOR LAND AND MARINE PURPOSES.**

By N. P. BURGH, Engineer. 12mo. . . . $2 00

**BURGH.—THE SLIDE-VALVE PRACTICALLY CONSIDERED:**

By N. P. BURGH, author of "A Treatise on Sugar Machinery," "Practical Illustrations of Land and Marine Engines," "A Pocket-Book of Practical Rules for Designing Land and Marine Engines, Boilers," etc. etc. etc. Completely illustrated. 12mo. . . . . . . . . . . . . $2 00

**BYRN.—THE COMPLETE PRACTICAL BREWER:**

Or, Plain, Accurate, and Thorough Instructions in the Art of Brewing Beer, Ale, Porter, including the Process of making Bavarian Beer, all the Small Beers, such as Root-beer, Ginger-pop, Sarsaparilla-beer, Mead, Spruce beer, etc. etc. Adapted to the use of Public Brewers and Private Families. By M. LA FAYETTE BYRN, M. D. With illustrations. 12mo. $1 25

**BYRN.—THE COMPLETE PRACTICAL DISTILLER:**

Comprising the most perfect and exact Theoretical and Practical Description of the Art of Distillation and Rectification; including all of the most recent improvements in distilling apparatus; instructions for preparing spirits from the numerous vegetables, fruits, etc.; directions for the distillation and preparation of all kinds of brandies and other spirits, spirituous and other compounds, etc. etc.; all of which is so simplified that it is adapted not only to the use of extensive distillers, but for every farmer, or others who may wish to engage in the art of distilling By M. LA FAYETTE BYRN, M. D. With numerous engravings. In one volume, 12mo. $1 50

**BYRNE.—POCKET BOOK FOR RAILROAD AND CIVIL ENGINEERS:**
Containing New, Exact, and Concise Methods for Laying out Railroad Curves, Switches, Frog Angles and Crossings; the Staking out of work; Levelling; the Calculation of Cuttings; Embankments; Earth-work, etc. By OLIVER BYRNE. Illustrated, 18mo. . . . . . . . . . $1 25

**BYRNE.—THE HANDBOOK FOR THE ARTISAN, MECHANIC, AND ENGINEER:**
By OLIVER BYRNE. Illustrated by 11 large plates and 185 Wood Engravings. 8vo. . . . . . . $5 00

**BYRNE.—THE ESSENTIAL ELEMENTS OF PRACTICAL MECHANICS:**
For Engineering Students, based on the Principle of Work. By OLIVER BYRNE. Illustrated by Numerous Wood Engravings, 12mo. . . . . . . . . . $3 63

**BYRNE.—THE PRACTICAL METAL-WORKER'S ASSISTANT:**
Comprising Metallurgic Chemistry; the Arts of Working all Metals and Alloys; Forging of Iron and Steel; Hardening and Tempering; Melting and Mixing; Casting and Founding; Works in Sheet Metal; the Processes Dependent on the Ductility of the Metals; Soldering; and the most Improved Processes and Tools employed by Metal-Workers. With the Application of the Art of Electro-Metallurgy to Manufacturing Processes; collected from Original Sources, and from the Works of Holtzapffel, Bergeron, Leupold, Plumier, Napier, and others. By OLIVER BYRNE. A New, Revised, and improved Edition, with Additions by John Scoffern, M. B , William Clay, Wm. Fairbairn, F. R. S., and James Napier. With Five Hundred and Ninety-two Engravings; Illustrating every Branch of the Subject. In one volume, 8vo. 652 pages . $7 00

**BYRNE.—THE PRACTICAL CALCULATOR:**
For the Engineer, Mechanic, Manufacturer of Engine Work, Naval Architect, Miner, and Millwright. By OLIVER BYRNE. 1 volume, 8vo., nearly 600 pages . . . . $4 50

**CABINET MAKER'S ALBUM OF FURNITURE:**
Comprising a Collection of Designs for the Newest and Most Elegant Styles of Furniture. Illustrated by Forty eight Large and Beautifully Engraved Plates. In one volume, oblong $5 00

**CALVERT.—LECTURES ON COAL-TAR COLORS, AND ON RECENT IMPROVEMENTS AND PROGRESS IN DYEING AND CALICO PRINTING:**

Embodying Copious Notes taken at the last London International Exhibition, and *Illustrated with Numerous Patterns of Aniline and other Colors.* By F. GRACE CALVERT, F. R. S., F. C. S., Professor of Chemistry at the Royal Institution, Manchester, Corresponding Member of the Royal Academies of Turin and Rouen; of the Pharmaceutical Society of Paris; Société Industrielle de Mulhouse, etc. In one volume, 8vo., cloth . . . . . . . . . $1 50

**CAMPIN.—A PRACTICAL TREATISE ON MECHANICAL ENGINEERING:**

Comprising Metallurgy, Moulding, Casting, Forging, Tools, Workshop Machinery, Mechanical Manipulation, Manufacture of Steam-engines, etc. etc. With an Appendix on the Analysis of Iron and Iron Ores. By FRANCIS CAMPIN, C. E. To which are added, Observations on the Construction of Steam Boilers, and Remarks upon Furnaces used for Smoke Prevention; with a Chapter on Explosions. By R. Armstrong, C. E., and John Bourne. Rules for Calculating the Change Wheels for Screws on a Turning Lathe, and for a Wheel-cutting Machine. By J. LA NICCA. Management of Steel, including Forging, Hardening, Tempering, Annealing, Shrinking, and Expansion. And the Case-hardening of Iron. By G. EDE. 8vo. Illustrated with 29 plates and 100 wood engravings. $6 00

**CAMPIN.—THE PRACTICE OF HAND-TURNING IN WOOD, IVORY, SHELL, ETC.:**

With Instructions for Turning such works in Metal as may be required in the Practice of Turning Wood, Ivory, etc. Also, an Appendix on Ornamental Turning. By FRANCIS CAMPIN; with Numerous Illustrations, 12mo., cloth . . . $3 00

**CAPRON DE DOLE.—DUSSAUCE.—BLUES AND CARMINES OF INDIGO.**

A Practical Treatise on the Fabrication of every Commercial Product derived from Indigo. By FELICIEN CAPRON DE DOLE. Translated, with important additions, by Professor H. DUSSAUCE. 12mo. . . . . . . . . . $2 50

CAREY.—THE WORKS OF HENRY C. CAREY:

CONTRACTION OR EXPANSION? REPUDIATION OR RESUMPTION? Letters to Hon. Hugh McCulloch. 8vo. 38

FINANCIAL CRISES, their Causes and Effects. 8vo. paper 25

HARMONY OF INTERESTS; Agricultural, Manufacturing, and Commercial. 8vo., paper . . . . . . $1 00
Do. do. cloth . . . . $1 50

LETTERS TO THE PRESIDENT OF THE UNITED STATES. Paper . . . . . . . . . . . 75

MANUAL OF SOCIAL SCIENCE. Condensed from Carey's "Principles of Social Science." By KATE MCKEAN. 1 vol. 12mo. . . . . . . . . . . . $2 25

MISCELLANEOUS WORKS: comprising "Harmony of Interests," "Money," "Letters to the President," "French and American Tariffs," "Financial Crises," "The Way to Outdo England without Fighting Her," "Resources of the Union," "The Public Debt," "Contraction or Expansion," "Review of the Decade 1857—'67," "Reconstruction," etc. etc. 1 vol. 8vo., cloth . . . . . . . . . . . $4 50

MONEY: A LECTURE before the N. Y. Geographical and Statistical Society. 8vo., paper . . . . . . 25

PAST, PRESENT, AND FUTURE. 8vo. . . . . $2 50

PRINCIPLES OF SOCIAL SCIENCE. 3 volumes 8vo., cloth $10 00

REVIEW OF THE DECADE 1857—'67. 8vo., paper 38

RECONSTRUCTION: INDUSTRIAL, FINANCIAL, AND POLITICAL. Letters to the Hon. Henry Wilson, U. S. S. 8vo. paper . . . . . . . . . 38

THE PUBLIC DEBT, LOCAL AND NATIONAL. How to provide for its discharge while lessening the burden of Taxation. Letter to David A. Wells, Esq., U. S. Revenue Commission. 8vo., paper . . . . . . . . . 25

THE RESOURCES OF THE UNION. A Lecture read, Dec. 1865, before the American Geographical and Statistical Society, N. Y., and before the American Association for the Advancement of Social Science, Boston . . . . 25

THE SLAVE TRADE, DOMESTIC AND FOREIGN; Why it Exists, and How it may be Extinguished. 12mo., cloth $1 50

THE WAY TO OUTDO ENGLAND WITHOUT FIGHTING HER. Letters to the Hon. Schuyler Colfax, Speaker of the House of Representatives United States, on "The Paper Question," "The Farmer's Question," "The Iron Question," "The Railroad Question," and "The Currency Question." 8vo., paper . . . . . . . . . . . . 75

**CHEVALIER.—THE PHOTOGRAPHIC STUDENT.**
A Complete Treatise on the Theory and Practice of Photography. Translated from the French of A. CHEVALIER. Illustrated by numerous engravings. (In press.)

**CLOUGH.—THE CONTRACTOR'S MANUAL AND BUILDER'S PRICE-BOOK:**
Designed to elucidate the method of ascertaining, correctly, the value and Quantity of every description of Work and Materials used in the Art of Building, from their Prime Cost in any part of the United States, collected from extensive experience and observation in Building and Designing; to which are added a large variety of Tables, Memoranda, etc., indispensable to all engaged or concerned in erecting buildings of any kind. By A. B. CLOUGH, Architect, 24mo., cloth 75

**COLBURN.—THE GAS-WORKS OF LONDON:**
Comprising a sketch of the Gas-works of the city, Process of Manufacture, Quantity Produced, Cost, Profit, etc. By ZERAH COLBURN. 8vo., cloth . . . . . . . . 75

**COLBURN.—THE LOCOMOTIVE ENGINE:**
Including a Description of its Structure, Rules for Estimating its Capabilities, and Practical Observations on its Construction and Management. By ZERAH COLBURN. Illustrated. A new edition. 12mo. . . . . . . . . $1 25

**COLBURN AND MAW.—THE WATER-WORKS OF LONDON:**
Together with a Series of Articles on various other Waterworks. By ZERAH COLBURN and W. MAW. Reprinted from "Engineering." In one volume, 8vo. . . $4 00

**DAGUERREOTYPIST AND PHOTOGRAPHER'S COMPANION:**
12mo., cloth . . . . . . . . . . . $1 25

**DAVIS.—A TREATISE ON HARNESS, SADDLES, AND BRIDLES:**
Their History and Manufacture from the Earliest Times down to the Present Period. By A. DAVIS, Practical Saddler and Harness Maker. (In press.)

**DESSOYE.—STEEL, ITS MANUFACTURE, PROPERTIES, AND USE.**

By J. B. J. Dessoye, Manufacturer of Steel; with an Introduction and Notes by Ed. Graten, Engineer of Mines. Translated from the French. In one volume, 12mo. (In press.)

**DIRCKS.—PERPETUAL MOTION:**

Or Search for Self-Motive Power during the 17th, 18th, and 19th centuries. Illustrated from various authentic sources in Papers, Essays, Letters, Paragraphs, and numerous Patent Specifications, with an Introductory Essay by Henry Dircks, C. E. Illustrated by numerous engravings of machines. 12mo., cloth . . . . . . . . . . $3 50

**DIXON.—THE PRACTICAL MILLWRIGHT'S AND ENGINEER'S GUIDE:**

Or Tables for Finding the Diameter and Power of Cogwheels; Diameter, Weight, and Power of Shafts; Diameter and Strength of Bolts, etc. etc. By Thomas Dixon. 12mo., cloth. $1 50

**DUNCAN.—PRACTICAL SURVEYOR'S GUIDE:**

Containing the necessary information to make any person, of common capacity, a finished land surveyor without the aid of a teacher. By Andrew Duncan. Illustrated. 12mo., cloth. $1 25

**DUSSAUCE.—A NEW AND COMPLETE TREATISE ON THE ARTS OF TANNING, CURRYING, AND LEATHER DRESSING:**

Comprising all the Discoveries and Improvements made in France, Great Britain, and the United States. Edited from Notes and Documents of Messrs. Sallerou, Grouvelle, Duval, Dessables, Labarraque, Payen, René, De Fontenelle, Malapeyre, etc. etc. By Prof. H. Dussauce, Chemist. Illustrated by 212 wood engravings. 8vo. . . . . . $10 00

**DUSSAUCE.—A GENERAL TREATISE ON THE MANUFACTURE OF EVERY DESCRIPTION OF SOAP:**

Comprising the Chemistry of the Art, with Remarks on Alkalies, Saponifiable Fatty Bodies, the apparatus necessary in a Soap Factory, Practical Instructions on the manufacture of the various kinds of Soap, the assay of Soaps, etc. etc. Edited from notes of Larmé, Fontenelle, Malapeyre, Dufour, and others, with large and important additions by Professor H. Dussauce, Chemist. Illustrated. In one volume, 8vo. (In press.)

**DUSSAUCE.—A PRACTICAL GUIDE FOR THE PERFUMER:**

Being a New Treatise on Perfumery the most favorable to the Beauty without being injurious to the Health, comprising a Description of the substances used in Perfumery, the Formulæ of more than one thousand Preparations, such as Cosmetics, Perfumed Oils, Tooth Powders, Waters, Extracts, Tinctures, Infusions, Vinaigres, Essential Oils, Pastels, Creams, Soaps, and many new Hygienic Products not hitherto described. Edited from Notes and Documents of Messrs. Debay, Lunel, etc. With additions by Professor H. DUSSAUCE, Chemist. (In press, *shortly to be issued.*)

**DUSSAUCE.—PRACTICAL TREATISE ON THE FABRICATION OF MATCHES, GUN COTTON, AND FULMINATING POWDERS.**

By Professor H. DUSSAUCE. 12mo. . . . . $3 00

**DUSSAUCE.—TREATISE ON THE COLORING MATTERS DERIVED FROM COAL TAR:**

Their Practical Application in Dyeing Cotton, Wool, and Silk; the Principles of the Art of Dyeing and of the Distillation of Coal Tar, with a Description of the most Important New Dyes now in use. By Prof. H. DUSSAUCE. 12mo. . $3 00

**DYER AND COLOR-MAKER'S COMPANION:**

Containing upwards of two hundred Receipts for making Colors, on the most approved principles, for all the various styles and fabrics now in existence; with the Scouring Process, and plain Directions for Preparing, Washing-off, and Finishing the Goods. In one vol. 12mo. . . . . . . $1 25

**EASTON.—A PRACTICAL TREATISE ON STREET OR HORSE-POWER RAILWAYS:**

Their Location, Construction, and Management; with General Plans and Rules for their Organization and Operation; together with Examinations as to their Comparative Advantages over the Omnibus System, and Inquiries as to their Value for Investment; including Copies of Municipal Ordinances relating thereto. By ALEXANDER EASTON, C. E. Illustrated by 23 plates, 8vo., cloth . . . . . . . . . $2 00

**ERNI.—COAL OIL AND PETROLEUM:**

Their Origin, History, Geology, and Chemistry; with a view of their importance in their bearing on National Industry. By Dr. HENRI ERNI, Chief Chemist, Department of Agriculture. 12mo. . . . . . . . . . . . . . $2 50

**ERNI.—THE THEORETICAL AND PRACTICAL CHEMISTRY OF FERMENTATION:**

Comprising the Chemistry of Wine, Beer, Distilling of Liquors; with the Practical Methods of their Chemical Examination, Preservation, and Improvement—such as Gallizing of Wines. With an Appendix, containing well-tested Practical Rules and Receipts for the manufacture, etc., of all kinds of Alcoholic Liquors. By HENRY ERNI, Chief Chemist, Department of Agriculture. (In press.)

**FAIRBAIRN.—THE PRINCIPLES OF MECHANISM AND MACHINERY OF TRANSMISSION:**

Comprising the Principles of Mechanism, Wheels, and Pulleys, Strength and Proportions of Shafts, Couplings of Shafts, and Engaging and Disengaging Gear. By WILLIAM FAIRBAIRN, Esq., C. E., LL. D., F. R. S., F. G. S., Corresponding Member of the National Institute of France, and of the Royal Academy of Turin; Chevalier of the Legion of Honor, etc. etc. Beautifully illustrated by over 150 wood-cuts. In one volume 12mo. $2 50

**FAIRBAIRN.—PRIME-MOVERS:**

Comprising the Accumulation of Water-power; the Construction of Water-wheels and Turbines; the Properties of Steam; the Varieties of Steam-engines and Boilers and Wind-mills. By WILLIAM FAIRBAIRN, C. E., LL. D., F. R. S., F. G. S. Author of "Principles of Mechanism and the Machinery of Transmission." With Numerous Illustrations. In one volume. (In press.)

**FLAMM.—A PRACTICAL GUIDE TO THE CONSTRUCTION OF ECONOMICAL HEATING APPLICATIONS FOR SOLID AND GASEOUS FUELS:**

With the Application of Concentrated Heat, and on Waste Heat, for the Use of Engineers, Architects, Stove and Furnace Makers, Manufacturers of Fire Brick, Zinc, Porcelain, Glass, Earthenware, Steel, Chemical Products, Sugar Refiners, Metallurgists, and all others employing Heat. By M. PIERRE FLAMM, Manufacturer. Illustrated. Translated from the French. One volume, 12mo. (In press.)

**GILBART.—A PRACTICAL TREATISE ON BANKING:**

By JAMES WILLIAM GILBART. To which is added: THE NATIONAL BANK ACT AS NOW (1868) IN FORCE. 8vo. $4 50

**GOTHIC ALBUM FOR CABINET MAKERS:**
Comprising a Collection of Designs for Gothic Furniture. Illustrated by twenty-three large and beautifully engraved plates. Oblong . . . . . . . . . $3 00

**GRANT.—BEET-ROOT SUGAR AND CULTIVATION OF THE BEET:**
By E. B. GRANT. 12mo. . . . . . . $1 25

**GREGORY.—MATHEMATICS FOR PRACTICAL MEN:**
Adapted to the Pursuits of Surveyors, Architects, Mechanics, and Civil Engineers. By OLINTHUS GREGORY. 8vo., plates, cloth . . . . . . . . . . . . $3 00

**GRISWOLD.—RAILROAD ENGINEER'S POCKET COMPANION.**
Comprising Rules for Calculating Deflection Distances and Angles, Tangential Distances and Angles, and all Necessary Tables for Engineers; also the art of Levelling from Preliminary Survey to the Construction of Railroads, intended Expressly for the Young Engineer, together with Numerous Valuable Rules and Examples. By W. GRISWOLD. 12mo., tucks. $1 25

**GUETTIER.—METALLIC ALLOYS:**
Being a Practical Guide to their Chemical and Physical Properties, their Preparation, Composition, and Uses. Translated from the French of A. GUETTIER, Engineer and Director of Founderies, author of "La Fouderie en France," etc. etc. By A. A. FESQUET, Chemist and Engineer. In one volume, 12mo. (In press, *shortly to be published.*)

**HATS AND FELTING:**
A Practical Treatise on their Manufacture. By a Practical Hatter. Illustrated by Drawings of Machinery, &c., 8vo.

**HAY.—THE INTERIOR DECORATOR:**
The Laws of Harmonious Coloring adapted to Interior Decorations: with a Practical Treatise on House-Painting. By D. R. HAY, House-Painter and Decorator. Illustrated by a Diagram of the Primary, Secondary, and Tertiary Colors. 12mo. $2 25

**HUGHES.—AMERICAN MILLER AND MILLWRIGHT'S ASSISTANT:**
By WM. CARTER HUGHES. A new edition. In one volume, 12mo. . . . . . . . . . $1 50

**HUNT.—THE PRACTICE OF PHOTOGRAPHY.**
By ROBERT HUNT, Vice-President of the Photographic Society, London, with numerous illustrations. 12mo., cloth . 75

**HURST.—A HAND-BOOK FOR ARCHITECTURAL SURVEYORS:**
Comprising Formulæ useful in Designing Builder's work, Table of Weights, of the materials used in Building, Memoranda connected with Builders' work, Mensuration, the Practice of Builders' Measurement, Contracts of Labor, Valuation of Property, Summary of the Practice in Dilapidation, etc. etc. By J. F. HURST, C. E. 2d edition, pocket-book form, full bound $2 50

**JERVIS.—RAILWAY PROPERTY:**
A Treatise on the Construction and Management of Railways; designed to afford useful knowledge, in the popular style, to the holders of this class of property; as well as Railway Managers, Officers, and Agents. By JOHN B. JERVIS, late Chief Engineer of the Hudson River Railroad, Croton Aqueduct, &c. One vol. 12mo., cloth . . . . . . . . $2 00

**JOHNSON.—A REPORT TO THE NAVY DEPARTMENT OF THE UNITED STATES ON AMERICAN COALS:**
Applicable to Steam Navigation and to other purposes. By WALTER R. JOHNSON. With numerous illustrations. 607 pp. 8vo., half morocco . . . . . . . . $6 00

**JOHNSON.—THE COAL TRADE OF BRITISH AMERICA:**
With Researches on the Characters and Practical Values of American and Foreign Coals. By WALTER R. JOHNSON, Civil and Mining Engineer and Chemist. 8vo. . . . $2 00

**JOHNSTON.—INSTRUCTIONS FOR THE ANALYSIS OF SOILS, LIMESTONES, AND MANURES.**
By J. W. F. JOHNSTON. 12mo. . . . . . 38

**KEENE.—A HAND-BOOK OF PRACTICAL GAUGING,**
For the Use of Beginners, to which is added A Chapter on Distillation, describing the process in operation at the Custom House for ascertaining the strength of wines. By JAMES B. KEENE, of H. M. Customs. 8vo. . . . . $1 25

**KENTISH.—A TREATISE ON A BOX OF INSTRUMENTS,**
And the Slide Rule; with the Theory of Trigonometry and Logarithms, including Practical Geometry, Surveying, Measuring of Timber, Cask and Malt Gauging, Heights, and Distances. By THOMAS KENTISH. In one volume. 12mo. . $1 25

**KOBELL.—ERNI.—MINERALOGY SIMPLIFIED:**

A short method of Determining and Classifying Minerals, by means of simple Chemical Experiments in the Wet Way. Translated from the last German Edition of F. Von Kobell, with an Introduction to Blowpipe Analysis and other additions. By Henri Erni, M. D., Chief Chemist, Department of Agriculture, author of "Coal Oil and Petroleum." In one volume, 12mo. . . . . . . . . $2 50

**LAFFINEUR.—A PRACTICAL GUIDE TO HYDRAULICS FOR TOWN AND COUNTRY;**

Or a Complete Treatise on the Building of Conduits for Water for Cities, Towns, Farms, Country Residences, Workshops, etc. Comprising the means necessary for obtaining at all times abundant supplies of Drinkable Water. Translated from the French of M. Jules Laffineur, C. E. Illustrated. (In press.)

**LAFFINEUR.—A TREATISE ON THE CONSTRUCTION OF WATER-WHEELS:**

Containing the various Systems in use with Practical Information on the Dimensions necessary for Shafts, Journals, Arms, etc., of Water-wheels, etc. etc. Translated from the French of M. Jules Laffineur, C. E. Illustrated by numerous plates. (In press.)

**LANDRIN.—A TREATISE ON STEEL:**

Comprising the Theory, Metallurgy, Practical Working, Properties, and Use. Translated from the French of H. C. Landrin, Jr., C. E. By A. A. Fesquet, Chemist and Engineer. Illustrated. 12mo. (In press.)

**LARKIN.—THE PRACTICAL BRASS AND IRON FOUNDER'S GUIDE:**

A Concise Treatise on Brass Founding, Moulding, the Metals and their Alloys, etc.; to which are added Recent Improvements in the Manufacture of Iron, Steel by the Bessemer Process, etc. etc. By James Larkin, late Conductor of the Brass Foundry Department in Reany, Neafie & Co.'s Penn Works, Philadelphia. Fifth edition, revised, with Extensive additions. In one volume, 12mo. . . . . . . $2 25

**LEAVITT.—FACTS ABOUT PEAT AS AN ARTICLE OF FUEL:**
With Remarks upon its Origin and Composition, the Localities in which it is found, the Methods of Preparation and Manufacture, and the various Uses to which it is applicable; together with many other matters of Practical and Scientific Interest. To which is added a chapter on the Utilization of Coal Dust with Peat for the Production of an Excellent Fuel at Moderate Cost, especially adapted for Steam Service. By H. T. LEAVITT. Third edition. 12mo. . . . . $1 75

**LEROUX.—A PRACTICAL TREATISE ON WOOLS AND WORSTEDS:**
Translated from the French of CHARLES LEROUX, Mechanical Engineer, and Superintendent of a Spinning Mill. Illustrated by 12 large plates and 34 engravings. In one volume 8vo. (In press, *shortly to be published.*)

**LESLIE (MISS).—COMPLETE COOKERY:**
Directions for Cookery in its Various Branches. By MISS LESLIE. 58th thousand. Thoroughly revised, with the addition of New Receipts. In 1 vol. 12mo., cloth . . $1 25

**LESLIE (MISS). LADIES' HOUSE BOOK:**
a Manual of Domestic Economy. 20th revised edition. 12mo., cloth . . . . . . . . . . . $1 25

**LESLIE (MISS).—TWO HUNDRED RECEIPTS IN FRENCH COOKERY.**
12mo. . . . . . . . . . . . 50

**LIEBER.—ASSAYER'S GUIDE:**
Or, Practical Directions to Assayers, Miners, and Smelters, for the Tests and Assays, by Heat and by Wet Processes, for the Ores of all the principal Metals, of Gold and Silver Coins and Alloys, and of Coal, etc. By OSCAR M. LIEBER. 12mo., cloth $1 25

**LOVE.—THE ART OF DYEING, CLEANING, SCOURING, AND FINISHING:**
On the most approved English and French methods; being Practical Instructions in Dyeing Silks, Woollens, and Cottons, Feathers, Chips, Straw, etc.; Scouring and Cleaning Bed and Window Curtains, Carpets, Rugs, etc.; French and English Cleaning, any Color or Fabric of Silk, Satin, or Damask. By THOMAS LOVE, a Working Dyer and Scourer. In 1 vol. 12mo. $3 00

**MAIN AND BROWN.—QUESTIONS ON SUBJECTS CONNECTED WITH THE MARINE STEAM-ENGINE:**

And Examination Papers; with Hints for their Solution. By THOMAS J. MAIN, Professor of Mathematics, Royal Naval College, and THOMAS BROWN, Chief Engineer, R. N. 12mo., cloth $1 50

**MAIN AND BROWN.—THE INDICATOR AND DYNAMOMETER:**

With their Practical Applications to the Steam-Engine. By THOMAS J. MAIN, M. A. F. R., Ass't Prof. Royal Naval College, Portsmouth, and THOMAS BROWN, Assoc. Inst. C. E., Chief Engineer, R. N., attached to the R. N. College. Illustrated. From the Fourth London Edition. 8vo. . . . $1 50

**MAIN AND BROWN.—THE MARINE STEAM-ENGINE.**

By THOMAS J. MAIN, F. R. Ass't S. Mathematical Professor at Royal Naval College, and THOMAS BROWN, Assoc. Inst. C. E. Chief Engineer, R. N. Attached to the Royal Naval College. Authors of "Questions connected with the Marine Steam-Engine," and the "Indicator and Dynamometer." With numerous Illustrations. In one volume, 8vo. . . . $5 00

**MAKINS.—A MANUAL OF METALLURGY:**

More particularly of the Precious Metals: including the Methods of Assaying them. Illustrated by upwards of 50 Engravings. By GEORGE HOGARTH MAKINS, M. R. C. S., F. C. S., one of the Assayers to the Bank of England, Assayer to the Anglo-Mexican Mints, and Lecturer upon Metallurgy at the Dental Hospital, London. In one volume, 12mo. . . $3 50

**MARTIN.—SCREW-CUTTING TABLES, FOR THE USE OF MECHANICAL ENGINEERS:**

Showing the Proper Arrangement of Wheels for Cutting the Threads of Screws of any required Pitch; with a Table for Making the Universal Gas-Pipe Thread and Taps. By W. A. MARTIN, Engineer. 8vo. . . . . . . 50

**MILES.—A PLAIN TREATISE ON HORSE-SHOEING.**

With illustrations. By WILLIAM MILES, author of "The Horse's Foot," . . . . . . . . . $1 00

**MOLESWORTH. POCKET-BOOK OF USEFUL FORMULÆ AND MEMORANDA FOR CIVIL AND MECHANICAL ENGINEERS.**

By GUILFORD L. MOLESWORTH, Member of the Institution of Civil Engineers, Chief Resident Engineer of the Ceylon Railway. Second American, from the Tenth London Edition. In one volume, full bound in pocket-book form . . $2 00

**MOORE.—THE INVENTOR'S GUIDE:**
Patent Office and Patent Laws; or, a Guide to Inventors, and a Book of Reference for Judges, Lawyers, Magistrates, and others. By J. G. MOORE. 12mo., cloth . . $1 25

**MOREAU.—PRACTICAL GUIDE FOR THE JEWELLER,**
In the Application of Harmony of Colors in the Arrangement of Precious Stones, Gold, etc., from the French of M. L. MOREAU, Jeweller and Designer. Illustrated. (In press.)

**NAPIER.—CHEMISTRY APPLIED TO DYEING.**
By JAMES NAPIER, F. C. S. A new and revised edition, brought down to the present condition of the Art. Illustrated. (In press.)

**NAPIER.—A MANUAL OF DYEING RECEIPTS FOR GENERAL USE.**
By JAMES NAPIER, F. C. S. *With Numerous Patterns of Dyed Cloth and Silk.* Second edition, revised and enlarged. 12mo. $3 75

**NAPIER.—MANUAL OF ELECTRO-METALLURGY:**
Including the Application of the Art to Manufacturing Processes. By JAMES NAPIER. Fourth American, from the Fourth London edition, revised and enlarged. Illustrated by engravings. In one volume, 8vo. . . . . . $2 00

**NEWBERY.—GLEANINGS FROM ORNAMENTAL ART OF EVERY STYLE;**
Drawn from Examples in the British, South Kensington, Indian, Crystal Palace, and other Museums, the Exhibitions of 1851 and 1862, and the best English and Foreign works. In a series of one hundred exquisitely drawn Plates, containing many hundred examples. By ROBERT NEWBERY. 4to. $15 00

**NICHOLSON.—A MANUAL OF THE ART OF BOOK-BINDING:**
Containing full instructions in the different Branches of Forwarding, Gilding, and Finishing. Also, the Art of Marbling Book-edges and Paper. By JAMES B. NICHOLSON. Illustrated. 12mo., cloth . . . . . . . $2 25

**NORRIS.—A HAND-BOOK FOR LOCOMOTIVE ENGINEERS AND MACHINISTS:**
Comprising the Proportions and Calculations for Constructing Locomotives; Manner of Setting Valves; Tables of Squares, Cubes, Areas, etc. etc. By SEPTIMUS NORRIS, Civil and Mechanical Engineer. New edition. Illustrated, 12mo., cloth $2 00

**NYSTROM.—ON TECHNOLOGICAL EDUCATION AND THE CONSTRUCTION OF SHIPS AND SCREW PROPELLERS:**

For Naval and Marine Engineers. By JOHN W. NYSTROM, late Acting Chief Engineer U. S. N. Second edition, revised with additional matter. Illustrated by seven engravings. 12mo. $2 50

**O'NEILL.—CHEMISTRY OF CALICO PRINTING, DYEING, AND BLEACHING:**

Including Silken, Woollen, and Mixed Goods; Practical and Theoretical. By CHARLES O'NEILL. (In press.)

**O'NEILL.—A DICTIONARY OF CALICO PRINTING AND DYEING:**

Containing a Brief Account of all the Substances and Processes in Use in the Arts of Printing and Dyeing Textile Fabrics; with Practical Receipts and Scientific Information. By CHARLES O'NEILL, Analytical Chemist, Fellow of the Chemical Society of London, etc. etc. Author of "Chemistry of Calico Printing and Dyeing." 8vo. (In press.)

**OVERMAN—OSBORN.—THE MANUFACTURE OF IRON IN ALL ITS BRANCHES:**

Including a Practical Description of the various Fuels and their Values, the Nature, Determination and Preparation of the Ore, the Erection and Management of Blast and other Furnaces, the characteristic results of Working by Charcoal, Coke, or Anthracite, the Conversion of the Crude into the various kinds of Wrought Iron, and the Methods adapted to this end. Also, a Description of Forge Hammers, Rolling Mills, Blast Engines, &c. &c. To which is added an Essay on the Manufacture of Steel. By FREDERICK OVERMAN, Mining Engineer. The whole thoroughly revised and enlarged, adapted to the latest Improvements and Discoveries, and the particular type of American Methods of Manufacture. With various new engravings illustrating the whole subject. By H. S. OSBORN, LL. D. Professor of Mining and Metallurgy in Lafayette College. In one volume, 8vo. (In press.) . $10 00

**PAINTER, GILDER, AND VARNISHER'S COMPANION:**

Containing Rules and Regulations in everything relating to the Arts of Painting, Gilding, Varnishing, and Glass Staining, with numerous useful and valuable Receipts; Tests for the Detection of Adulterations in Oils and Colors, and a statement of the Diseases and Accidents to which Painters, Gilders, and

Varnishers are particularly liable, with the simplest methods of Prevention and Remedy. With Directions for Graining. Marbling, Sign Writing, and Gilding on Glass. To which are added COMPLETE INSTRUCTIONS FOR COACH PAINTING AND VARNISHING. 12mo., cloth . . . . . $1 50

**PALLETT.—THE MILLER'S, MILLWRIGHT'S, AND ENGINEER'S GUIDE.**

By HENRY PALLETT. Illustrated. In one vol. 12mo. $3 00

**PERKINS.—GAS AND VENTILATION.**

Practical Treatise on Gas and Ventilation. With Special Relation to Illuminating, Heating, and Cooking by Gas. Including Scientific Helps to Engineer-students and others. With illustrated Diagrams. By E. E. PERKINS. 12mo., cloth $1 25

**PERKINS AND STOWE.—A NEW GUIDE TO THE SHEET-IRON AND BOILER PLATE ROLLER:**

Containing a Series of Tables showing the Weight of Slabs and Piles to Produce Boiler Plates, and of the Weight of Piles and the Sizes of Bars to produce Sheet-iron; the Thickness of the Bar Gauge in Decimals; the Weight per foot, and the Thickness on the Bar or Wire Gauge of the fractional parts of an inch; the Weight per sheet, and the Thickness on the Wire Gauge of Sheet-iron of various dimensions to weigh 112 lbs. per bundle; and the conversion of Short Weight into Long Weight, and Long Weight into Short. Estimated and collected by G. H. PERKINS and J. G. STOWE . . . . . $2 50

**PHILLIPS AND DARLINGTON.—RECORDS OF MINING AND METALLURGY:**

Or Facts and Memoranda for the use of the Mine Agent and Smelter. By J. ARTHUR PHILLIPS, Mining Engineer, Graduate of the Imperial School of Mines, France, etc., and JOHN DARLINGTON. Illustrated by numerous engravings. In one volume, 12mo. . . . . . . . . . $2 00

**PRADAL, MALEPEYRE, AND DUSSAUCE. — A COMPLETE TREATISE ON PERFUMERY:**

Containing notices of the Raw Material used in the Art, and the Best Formulæ. According to the most approved Methods followed in France, England, and the United States. By M. P. PRADAL, Perfumer Chemist, and M. F. MALEPEYRE. Translated from the French, with extensive additions, by Professor H. DUSSAUCE. 8vo. . . . . . . . . $10 00

**PROTEAUX.—PRACTICAL GUIDE FOR THE MANUFACTURE OF PAPER AND BOARDS.**

By A. PROTEAUX, Civil Engineer, and Graduate of the School of Arts and Manufactures, Director of Thiers's Paper Mill, 'Puy-de-Dômé. With additions, by L. S. LE NORMAND. Translated from the French, with Notes, by HORATIO PAINE, A. B., M. D. To which is added a Chapter on the Manufacture of Paper from Wood in the United States, by HENRY T. BROWN, of the "American Artisan." Illustrated by six plates, containing Drawings of Raw Materials, Machinery, Plans of Paper-Mills, etc. etc. 8vo. . . . . . . $5 00

**REGNAULT.—ELEMENTS OF CHEMISTRY.**

By M. V. REGNAULT. Translated from the French by T. FORREST BETTON, M. D., and edited, with notes, by JAMES C. BOOTH, Melter and Refiner U. S. Mint, and WM. L. FABER, Metallurgist and Mining Engineer. Illustrated by nearly 700 wood engravings. Comprising nearly 1500 pages. In two volumes, 8vo., cloth . . . . . . . $10 00

**SELLERS.—THE COLOR MIXER:**

Containing nearly Four Hundred Receipts for Colors, Pastes, Acids, Pulps, Blue Vats, Liquors, etc. etc., for Cotton and Woollen Goods: including the celebrated Barrow Delaine Colors. By JOHN SELLERS, an experienced Practical Workman. In one volume, 12mo. . . . . . . $2 50

**SHUNK.—A PRACTICAL TREATISE ON RAILWAY CURVES AND LOCATION, FOR YOUNG ENGINEERS.**

By WM. F. SHUNK, Civil Engineer. 12mo. . . . $1 50

**SMEATON.—BUILDER'S POCKET COMPANION:**

Containing the Elements of Building, Surveying, and Architecture; with Practical Rules and Instructions connected with the subject. By A. C. SMEATON, Civil Engineer, etc. In one volume, 12mo. . . . . . . . . $1 25

**SMITH.—THE DYER'S INSTRUCTOR:**

Comprising Practical Instructions in the Art of Dyeing Silk, Cotton, Wool, and Worsted, and Woollen Goods: containing nearly 800 Receipts. To which is added a Treatise on the Art of Padding; and the Printing of Silk Warps, Skeins, and Handkerchiefs, and the various Mordants and Colors for the different styles of such work. By DAVID SMITH, Pattern Dyer. 12mo., cloth. . . . . . . . $3 00

**SMITH.—PARKS AND PLEASURE GROUNDS:**

Or Practical Notes on Country Residences, Villas, Public Parks, and Gardens. By CHARLES H. J. SMITH, Landscape Gardener and Garden Architect, etc. etc. 12mo. . $2 25

**STOKES.—CABINET-MAKER'S AND UPHOLSTERER'S COMPANION:**

Comprising the Rudiments and Principles of Cabinet-making and Upholstery, with Familiar Instructions, Illustrated by Examples for attaining a Proficiency in the Art of Drawing, as applicable to Cabinet-work; The Processes of Veneering, Inlaying, and Buhl-work; the Art of Dyeing and Staining Wood, Bone, Tortoise Shell, etc. Directions for Lackering, Japanning, and Varnishing; to make French Polish; to prepare the Best Glues, Cements, and Compositions, and a number of Receipts particularly for workmen generally. By J. STOKES. In one vol. 12mo. With illustrations . . . . . $1 25

**STRENGTH AND OTHER PROPERTIES OF METALS.**

Reports of Experiments on the Strength and other Properties of Metals for Cannon. With a Description of the Machines for Testing Metals, and of the Classification of Cannon in service. By Officers of the Ordnance Department U. S. Army. By authority of the Secretary of War. Illustrated by 25 large steel plates. In 1 vol. quarto . . . . . . $10 00

**TABLES SHOWING THE WEIGHT OF ROUND, SQUARE, AND FLAT BAR IRON, STEEL, ETC.,**

By Measurement. Cloth . . . . . . . 63

**TAYLOR.—STATISTICS OF COAL:**

Including Mineral Bituminous Substances employed in Arts and Manufactures; with their Geographical, Geological, and Commercial Distribution and amount of Production and Consumption on the American Continent. With Incidental Statistics of the Iron Manufacture. By R. C. TAYLOR. Second edition, revised by S. S. HALDEMAN. Illustrated by five Maps and many wood engravings. 8vo., cloth . . . . $6 00

**TEMPLETON.—THE PRACTICAL EXAMINATOR ON STEAM AND THE STEAM-ENGINE:**

With Instructive References relative thereto, for the Use of Engineers, Students, and others. By WM. TEMPLETON, Engineer. 12mo. . . . . . . . . . $1 25

**THOMAS.—THE MODERN PRACTICE OF PHOTOGRAPHY.**
By R. W. THOMAS, F. C. S. 8vo., cloth . . . . 75

**THOMSON.—FREIGHT CHARGES CALCULATOR.**
By ANDREW THOMSON, Freight Agent . . . . $1 25

**TURNBULL.—THE ELECTRO-MAGNETIC TELEGRAPH:**
With an Historical Account of its Rise, Progress, and Present Condition. Also, Practical Suggestions in regard to Insulation and Protection from the effects of Lightning. Together with an Appendix, containing several important Telegraphic Devices and Laws. By LAWRENCE TNRNBULL, M. D., Lecturer on Technical Chemistry at the Franklin Institute. Revised and improved. Illustrated. 8vo. . . . . $3 00

**TURNER'S (THE) COMPANION:**
Containing Instructions in Concentric, Elliptic, and Eccentric Turning; also various Plates of Chucks, Tools, and Instruments; and Directions for using the Eccentric Cutter, Drill, Vertical Cutter, and Circular Rest; with Patterns and Instructions for working them. A new edition in one vol. 12mo.
$1 50

**ULRICH—DUSSAUCE.—A COMPLETE TREATISE ON THE ART OF DYEING COTTON AND WOOL:**
As practised in Paris, Rouen, Mulhausen, and Germany. From the French of M. LOUIS ULRICH, a Practical Dyer in the principal Manufactories of Paris, Rouen, Mulhausen, etc. etc.; to which are added the most important Receipts for Dyeing Wool, as practised in the Manufacture Impériale des Gobelins, Paris. By Professor H. DUSSAUCE. 12mo. $3 00

**URBIN—BRULL.—A PRACTICAL GUIDE FOR PUDDLING IRON AND STEEL.**
By ED. URBIN, Engineer of Arts and Manufactures. A Prize Essay read before the Association of Engineers, Graduate of the School of Mines, of Liege, Belgium, at the Meeting of 1865—6. To which is added a COMPARISON OF THE RESISTING PROPERTIES OF IRON AND STEEL. By A. BRULL. Translated from the French by A. A. FESQUET, Chemist and Engineer. In one volume, 8vo. . . . . . . . . . $1 00

**WATSON.—A MANUAL OF THE HAND-LATHE.**
By EGBERT P. WATSON, Late of the "Scientific American," Author of "Modern Practice of American Machinists and Engineers." In one volume, 12mo. (In press.)

**WATSON.—THE MODERN PRACTICE OF AMERICAN MACHINISTS AND ENGINEERS:**
Including the Construction, Application, and Use of Drills, Lathe Tools, Cutters for Boring Cylinders, and Hollow Work Generally, with the most Economical Speed of the same, the Results verified by Actual Practice at the Lathe, the Vice, and on the Floor. Together with Workshop management, Economy of Manufacture, the Steam-Engine, Boilers, Gears, Belting, etc. etc. By EGBERT P. WATSON, late of the "Scientific American." Illustrated by eighty-six engravings. 12mo. . . $2 50

**WATSON.—THE THEORY AND PRACTICE OF THE ART OF WEAVING BY HAND AND POWER:**
With Calculations and Tables for the use of those connected with the Trade. By JOHN WATSON, Manufacturer and Practical Machine Maker. Illustrated by large drawings of the best Power-Looms. 8vo. . . . . . . $7 50

**WEATHERLY.—TREATISE ON THE ART OF BOILING SUGAR, CRYSTALLIZING, LOZENGE-MAKING, COMFITS, GUM GOODS,**
And other processes for Confectionery, &c. In which are explained, in an easy and familiar manner, the various Methods of Manufacturing every description of Raw and Refined sugar Goods, as sold by Confectioners and others . . $2 00

**WILL.—TABLES FOR QUALITATIVE CHEMICAL ANALYSIS.**
By Prof. HEINRICH WILL, of Giessen, Germany. Seventh edition. Translated by CHARLES F. HIMES, Ph. D., Professor of Natural Science, Dickinson College, Carlisle, Pa. . $1 25

**WILLIAMS.—ON HEAT AND STEAM:**
Embracing New Views of Vaporization, Condensation, and Expansion. By CHARLES WYE WILLIAMS, A. I. C. E. Illustrated. 8vo. . . . . . . . . . $3 50